SHOVEL-TRUCK SYSTEMS

Shovel-Truck Systems

Modelling, Analysis and Calculation

Jacek M. Czaplicki
Mining Mechanization Institute, Silesian University of Technology, Gliwice, Poland

CRC Press
Taylor & Francis Group
Boca Raton London New York Leiden

CRC Press is an imprint of the
Taylor & Francis Group, an **informa** business
A BALKEMA BOOK

CRC Press/Balkema is an imprint of the Taylor & Francis Group, an informa business

Typeset by Vikatan Publishing Solutions (P) Ltd., Chennai, India.
Printed and bound in Great Britain by Antony Rowe (a CPI Group company), Chippenham, Wiltshire.

Published by: CRC Press/Balkema
P.O. Box 447, 2300 AK Leiden, The Netherlands
e-mail: Pub.NL@taylorandfrancis.com
www.crcpress.com – www.taylorandfrancis.co.uk – www.balkema.nl

ISBN: 978-0-415-48135-9 (hbk)
ISBN: 978-0-203-88124-8 (ebook)

Contents

Preface and acknowledgements

Authors of books often try to trace the point at which their work was first conceived. In the case of this monograph, it seems to me that it was in 1987, when I stood in front of the Nchanga Open Pit, Zambia and observed the machinery system involved. For the first time I saw a large open pit mine and huge machines in operation. At that point I had been a university lecturer, with a mine mechanization specialization, for some fifteen years, and immediately many questions came to mind, such as: *How many trucks should be in operation? How many trucks should be in reserve? How can the whole system be computed?* I tried to find answers to these problems, but many of them remained unknown. Many solutions that I found in the literature generated doubts, so I made the decision to carry out my own research work in this field. I returned to this problem many times over the next fifteen years and finally constructed a model describing the operation of the shovel-truck system that allowed the basic system parameters to be calculated taking into account most of the stochastic phenomena that occurred during operation.

In 2004, my textbook written for Polish students summarized the problems connected with cyclic machinery systems for mining and earthmoving—problems that can be considered in the field of queue theory. Two years later, I wrote a monograph on the shovel-truck system exclusively for the purposes of acquiring my last academic scientific degree. This monograph, in turn, is a general review of my last book with many necessary modifications and a significant extension of the considerations in this field. The 2006 monograph's main point of consideration was modelling alone. This book tackles analysis and calculations, taking into account for the first time the spare loader problem and priority during truck dispatching.

This paper is directed to the students of mining faculties and schools of mines with a specialization in mine mechanization all over the world. Some parts should be interesting for those who specialize in earthmoving enterprises. Students of mathematics searching for a practical application of queue theory models will also find some chapters interesting.

In Poland there has been a strong mining industry for more than a hundred years, but no open pits. Therefore, shovel-truck systems are rarely applied in mining and the size of equipment is small. For this reason, I feel that the results of my research work should be published in English to allow the people involved in the operation of shovel-truck systems to make use of these research results. They should be the main practical beneficiaries of it. I hope that my academic colleagues working at mining universities will also find this book interesting, and perhaps useful in their educational work. They too can verify how good the procedure is.

I would like to express my very warm thanks to Janjaap Blom, Senior Publisher for Taylor & Francis for his reliable and efficient cooperation.

Jacek M. Czaplicki
Mining Mechanization Institute
Silesian University of Technology, Gliwice, Poland

About the author

Jacek M. Czaplicki has been an academic lecturer for more than thirty years and is continuously associated with his home University. He did however leave his school for a couple of years' lecturing in African universities.

He worked for three years at the Kwara State College of Technology, Ilorin, Nigeria on a UNESCO project. A few years later he was appointed to Zambia Consolidated Copper Mines Ltd and worked as a lecturer at the School of Mines of the University of Zambia as part of a World Bank project.

Czaplicki received a Master of Science in Mine Mechanization from the Silesian University of Technology, Gliwice, Poland. He also obtained a Doctorate degree in Technical Sciences. Later he submitted a thesis and passed all requirements, obtaining a D. Sc. degree in Mining and Geological Engineering with a specialization in Mine Machinery at the same home University.

He has published more than a hundred and twenty papers in Poland and abroad. His specialization comprises mine transport, reliability and computation of mine machinery and their systems and reliability of hoist head ropes. He is an internationally recognized specialist in mine mechanization.

List of major notations

A – steady-state availability
A_k – steady-state availability of shovel
A_n – steady-state availability of repair station
A_{ts} – steady-state availability of truck in system
A_w – steady-state availability of truck
A'_w – adjusted steady-state availability of truck
$b \in \mathfrak{R}$ – current indicator determining number of trucks in the shovel-truck system
B_k – accessibility coefficient of shovel
c – normalization constant
C_M – square of coefficient of variation of work times
C_R – square of coefficient of variation of repair times
C_S – square of coefficient of variation of times to failure for machine in reserve
C_1 – square of coefficient of variation of truck travel times
C_2 – square of coefficient of variation of loading times
$d \in \mathfrak{R}$ – current indicator determining number of shovels
$\boldsymbol{D}$ – random variable, number of machines in work state
E – expected value, mathematical hope
E_p – expected value of number of machines in work state
E_u – expected value of number of failed machines
$f(x,b)$ – probability density function of number of trucks at loading shovels
$g(x)$ – component function of probability density function $f(x,b)$
G – general distribution
G_k – shovel loading capability coefficient
h – required number of trucks in system
$h(x)$ – component function of probability density function,
$i \in \mathfrak{R}$ – current indicator determining number of repair stands
$j \in \mathfrak{R}$ – current indicator determining number of failed trucks
k – number of service stands
K – constant, component of probability density function $f(x,b)$; function in randomised models
m – number of trucks directed to accomplish transportation task
n – number of shovels in system
M – exponential distribution

O – mean haulage time of truck

P, p – probability

p_q – probability of loading of g trucks

p_{gd} – conditional probability of event that there will be g trucks at d shovels able to load

$p_{g(d+s)}$ – conditional probability of event that there will be g trucks at d shovels able to load and s front-end loaders able to load, $d+s=n$

$p'_{g(d+s)}$ – conditional probability of event that there will be g trucks at d shovels able to load and s front-end loaders able to load, $d+s=n$; this probability concerns the system after change in its organization

P_j – probability that j machines fail

Q – truck loading capacity

R – mean return time of truck from dumping point to shovel

r – reserve size, number of spare trucks

t – time

T_c – mean time of truck work cycle

T_j – mean time of truck travel (haulage + dump + return)

T_n, T – mean time: repair, work

$T_{ow'}$ – truck mean time waiting for repair

T_{ns} – mean time of state of truck unserviceability

u_t – transport truck rate

V – number of trucks needed

W – mean dump time of truck

W_{efk} – effective output of shovel system

W_{efw} – effective output of truck system

W_{pk} – potential output of shovel system

W_{pw} – output of truck system

W_{tk} – theoretical output of shovel system

W_{wk} – specific output of shovel system

$\boldsymbol{X}$ – random variable, number of failed trucks

z_t – transportation task in unit of time

Z, Z' – mean time and adjusted mean time of loading

α – intensity of failures machines in reserve

β – power exponent

γ – intensity of truck repair

δ – intensity of truck failures

Δ – time loss function for truck

ε – intensity of shovel failures

ϖ – relative intensity coefficient of truck loading

η – repair intensity of shovel

θ – failure rate of machines in reserve

κ – failure rate (fault coefficient)

Λ – intensity of arrivals to repair system

ξ – failure rate of truck in work state

θ – mean truck queue length

ρ – flow intensity rate in service system

σ – standard deviation

σ_n – standard deviation of repair times

σ_p – standard deviation of work times

τ – proportional coefficient indicating how many times longer the mean time of truck loading by front-end loader is compared to the adjusted mean loading time by shovel

Ψ – coefficient regarding continuity of density function $f(x,b)$

$\mathfrak{H}$ – set of shovels

$\mathfrak{K}_{np}$ – repair state of shovel

$\mathfrak{K}_{nd}$ – shovel state of inaccessibility for loading

$\mathfrak{K}_{nz}$ – shovel state of incapability for loading

$\mathfrak{K}_p$ – work state of shovel

$\mathfrak{K}_{zd}$ – shovel state of capability for loading

$\mathfrak{N}$ – set of natural numbers

$\mathfrak{S}$ – system

$\mathfrak{T}$ – truck type

$\mathfrak{W}$ – set of trucks

$\mathfrak{Z}$ – principle to keep constant proportion of number of trucks to accomplish transportation task to number of shovels able to load

Random variables are marked in bold; this does not apply to gothic letters.

CHAPTER 1

Introduction

World mining today is at a particularly important stage in its development. Such a statement can be made after analyzing papers issued by the British Geological Survey and Natural Environment Research Council (Taylor *et al.* 2006). This phase is determined by several processes, such as:

- Continuous increment in the tendency of prices of the majority of mineral commodities
- Continuous booming demand for heavy mining equipment and permanent growth in production of large machinery for this type of industry
- The strengthened position of surface mining in mining production as a whole.

The prices of mineral commodities have been growing for a number of years.[1] The world market, especially the Chinese portion, calls for greater delivery of the fruits of mine production. Metallic ores are particularly wanted by developing countries (Weber 2005, White 2006). All of these factors mean that the extraction of minerals is still a good business, in spite of the fact that exploitation conditions are growing continuously more disadvantageous. So-called *easy deposits* have been depleted; now is the time to reach deposits lying deeper or in more remote, difficult areas. Mines have more money to improve their production, to take greater care over safety issues, and to look more carefully for the novel technical solutions offered by universities and research centres in cooperation with manufacturers.

It is a well-known fact that surface mining delivers the majority of mineral production, as far as the mass of this production is concerned. Some researchers are of the opinion that it accounts for approximately 90% of the total production in this field. The majority of metal ores are extracted by surface methods. For these reasons, during the last few years machinery producers have obtained orders at record-breaking levels. (www.jsonline.com, www.allmining.com, www.prweb.com, Gilewicz and Woof 2005). The problems connected with surface mining production are becoming more significant. The correct arrangement of machinery systems in mines and control of their operation are crucial dilemmas. However, to improve mine output and to make it more beneficial, all the mechanisms that have an influence on the course of the operation processes running in the mine need to be known. Apart from deterministic phenomena, many stochastic courses of actions can be observed during the exploitation of mineral deposits. To identify them, to find mutual relationships between them and to model them in such a way that analysis, inference and later decision-making can be made properly is often difficult, sometimes very difficult. Fortunately, it is not a *mission impossible*. This book aims to prove the above two statements.

1.1 OPEN PIT MINES

Generally, the extraction of mineral deposits from the surface by entry methods characterizes four types of mines. These are:

- Alluvial mines
- Quarries
- Open cast or strip mines
- Open pit mines.

This division is connected with both the kind of deposit extracted and the mining method applied.

---

[1] These considerations were made before autumn 2008 financial collapse at Wall Street that caused fall down of some mineral commodities.

Alluvial mining concerns working loose unconsolidated deposits (placer deposits) and extraction of such minerals as: diamond, gold, titanium, wolfram, tin and platinum. It mainly involves the application of such equipment as dredges and monitors. However, pans, sluice boxes, trommels, etc. should be encountered here as well as the bucket wheel excavators, draglines, dozers, hoppers, etc. employed in dry mining in Australia.

Quarry mining connected with the working of massive type deposits and chemical stones is usually divided into two categories:

- Dimension stone mining
- Broken stone mining and limestone extraction.

The equipment used here is quite different. Starting from a variety of saws, water and flame jets, drillers, cranes, derricks, hoists, trucks, etc. employed in dimension stone, it works up to the crushers, sizers, screens, front-end loaders and dozers applied in broken stone enterprises.

Open cast mining concerns bedded type deposits. However, lenticular or pocket deposits and massive deposits can in some cases be extracted by this method. The size of mine is usually quite large, up to twenty kilometres square in cross-section area, and the speed of development is high. Due to the application of continuous operation machinery—bucket wheel excavators, bucket chain excavators, surface continuous miners, as well as belt conveyors, stackers, etc.—this type of mining is termed 'continuous'. Apart from the first stage of mine development, the removed overburden is always located in the mined part of the works. Bituminous sands mining can be encountered in open cast mining.

Open pit mines are usually associated with vein type deposits, but this type of working can also sometimes be found in the exploitation of lenticular or massive type deposits. In special cases, it can be applied in the extraction of bedded deposits.

There are two main characteristic features of these types of workings:

- The manner of waste location and scale of this operation, and
- The spatial shape of the mine.

One of the main characteristic features of open pit mining is the almost continuous process of waste/overburden location outside the works. The depth of pit usually increases with the period of mine development and the shape of the mine is similar to a converted irregular cone (e.g. Rössing Uranium, Namibia), also sometimes called a funnel shape. However, in some cases this shape is surprisingly regular (e.g. Bingham Canyon, USA).

Both unit operations running in an open pit mine, i.e. overburden removal and extraction of mineral, are made in the same way—drilling and blasting. Practically the only factor that determines this is the property of the rock being excavated. Usually, this rock is an old geological formation that has been in the earth for such a long time that it is tough and hard to excavate. Blasting must be done in such a way that the transport road is not blocked, even for a short period of time. Mine planning must be done in the proper way. The removed overburden must expose the mineral deposit promptly, allowing for its extraction on an appropriate scale. The dimensions of benches, transporting routes and working rooms are determined by the size of machinery applied. The slope angle is determined by safety regulations. The whole operation runs until the moment when the cut-off point is reached. Further exploitation becomes economically impractical, though the underground mining operation of such deposits can be conducted. The moment at which this point occurs depends on many factors, such as current and future mineral price, mining and geological conditions of the deposit and surrounding rocks, and the machinery equipment involved. In the 1950s the depth of surface mines was counted as 300 to 400 m. Currently, Chuquicamata in Chile has a depth of approximately 950 m and extraction is still taking place. The planned depth should reach 1100 m two years from now.

1.2 MACHINERY SYSTEMS APPLIED

Another characteristic feature of open pit mining is the machinery system employed to realize the extraction of rock and loading and hauling of material. This material, especially in the case of vein type deposits, possesses very inconvenient properties for both extraction and transportation. The material, which is billions of years old, has been compressed by different pressures, displaced by different forces, tectonic movements, activated by different temperatures, chemicals and other processes, is massive, tough, hard to excavate, characterized by sharp edges and frequently abrasive. The only economical way to extract it is by drilling and blasting. This blasted material still has unfavourable properties for loading machines and hauling means. The only way is to load it by buckets or dippers into haulage boxes fixed to a means of transport. Frequently pieces of blasted rock are big and difficult to load, even by large buckets. Sometimes secondary blasting is inevitable.

The sequence of works is as follows—Figure 1.1.

At first, the places where the explosives will be loaded and then fired are determined. This concerns the blasting of both overburden and mineral. The Overburden must be removed in such a way that the extraction of the mineral can continue without interruption. To accomplish this preliminary operation, drillers are applied and loading equipment used to locate explosives in drilled holes. During blasting, the majority of operations in the mine are stopped. Loading is then performed by power shovels, and sometimes additionally by front-end loaders. These huge loading machines are the only ones that can cope successfully with this unfavourable broken rock. The capacity of buckets reaches 180 ton, although machines cooperating with regular hauling trucks have smaller buckets—up to 100 or 120 ton. The prices of these loading units reach several dozen million US$. Their durability usually reaches 20–30 years, and the total mass up to more than a hundred ton. Annual hours of use amount to approximately 7000. The steady-state availability usually varies between 0.80 and 0.90 for good quality machines. The time taken for material to be located from one bucket onto a truck is below 1 minute, and for a truck to be loaded usually about 2 to 3 minutes. The minimum number of buckets to fill a truck box is actually 3. Two buckets can also be applied, but if the bucket capacity is large enough to load a truck in 2 or 3 passes some disadvantageous effects may be inevitable. The first is the high possibility of frequent spillage of material around the truck being loaded. This leads to a longer duration of the truck work cycle because of the necessity of clean-up procedures in the loading area. The second factor is the lower durability of some truck assemblies due to frequent, high-impact forces acting on the vehicle during loading. The lifetime of the box lining can also be reduced. Great power shovels require such an amount of money that they should be in operation 24 hours a day, 7 days a week. These huge units are either of the 'rope' unit is mechanical; but hydraulic machines of this kind are still growing and masses of 80–90 ton can currently be accommodated in their bucket. Hydraulic shovels have several significant merits compared to mechanical machines, so the hard competition is still on.

Blasted rock is usually loaded onto hauling trucks, although some modifications to this type of system can be observed. The transported waste material is located on dumps waiting for the end of mining operations. Later it is used to fill up the great hole that remains after exploitation. The transported minerals are delivered to the peripheral equipment of the dressing plant.

Several decades ago some mines around the world applied rail haulage. A key factor that has to be taken into account in terms of rail transportation is the necessity to assure almost entirely horizontal track placement. This condition has to a great extent restricted opportunities to employ this type of haulage. This kind of system was applied in Bingham Canyon. Boxes of wagons are well prepared for transportation of such inconvenient material and generally trains are reliable. However, the application of this type of transportation only makes economic sense if the distance travelled is appropriately long. In the engineering of rail transportation it is clearly stated: 'huge masses, long distance, long life of system, horizontal track'. Therefore, the number of rail systems in open-pit mines in the world is rather small, currently amounting to just a few percent of the total number of machinery systems applied.

The application of trucks in open pit mining is common, despite of the fact that they have several significant disadvantages. On average, this kind of vehicle consumes about 60% of its energy to move itself. Approximately 40% is connected with displacement of the payload. Compared to a conventional belt conveyor that uses about 20% of its energy to move itself and

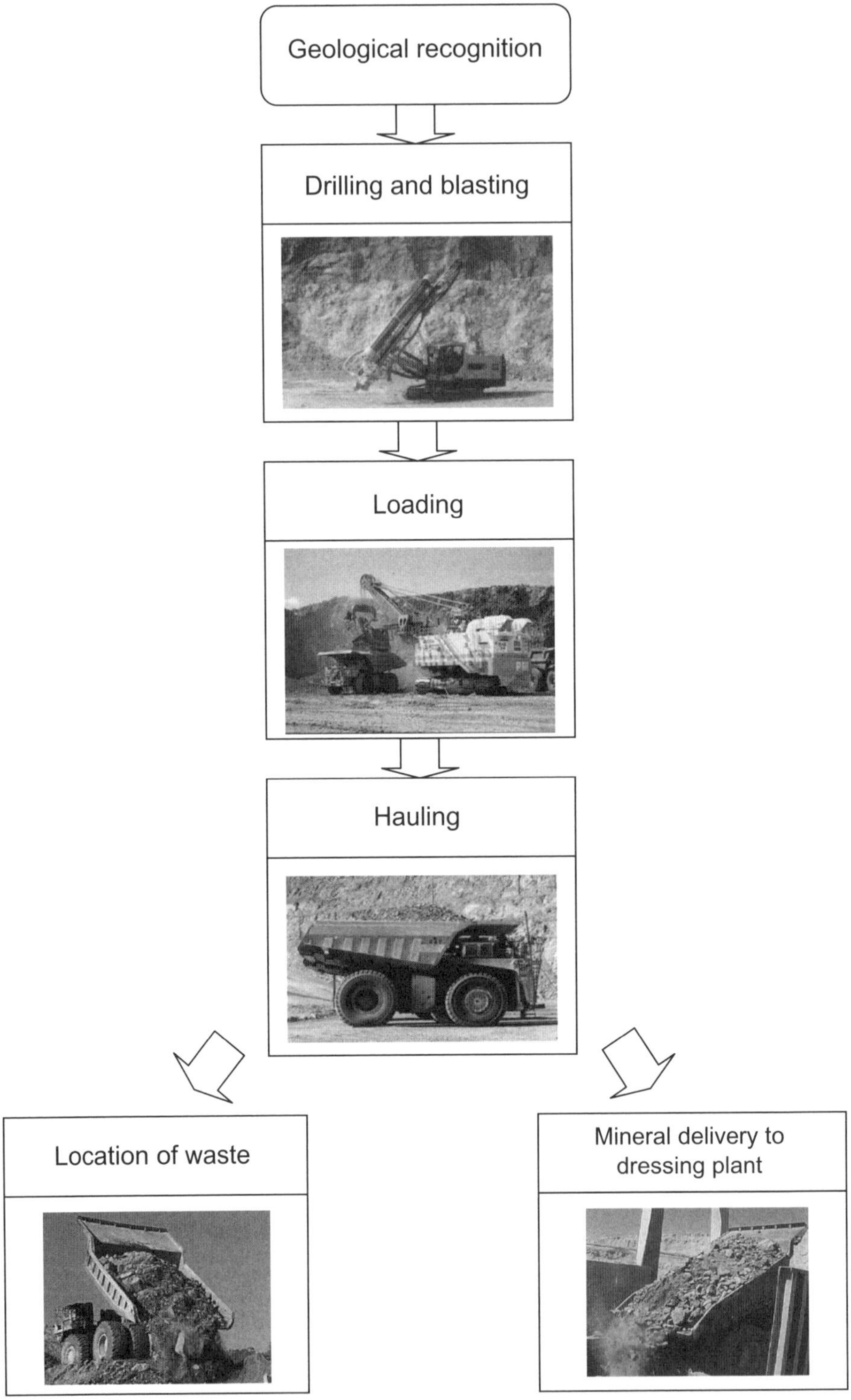

Figure 1.1. Sequence of main works in open pit.

a further 80% to transport bulk, the parameters of the truck are not favourable. Notice that 50% of the route truck moves empty. It consumes a huge quantity of fuel moving at high speed, say at 10 km/h, up the pit ramp. The total mass in motion actually reaches 600 ton. An economical travel distance for dumpers is only a few kilometres. For large systems employing huge trucks, it is compulsory to build several monster fuel tanks, as well as a large maintenance bay. The application of trolley assist systems is frequently profitable, especially in countries where the price of electrical energy is low. Trucks have two very convenient features. They can serve many different points, even if these points change their position frequently. Additionally, if the truck fails, the driver changes his seat onto a unit taken from the reserve. The whole system is still working.

Truck technology has developed over the last few decades in a way that is probably the most spectacular among mine machines, and progress continues to take place. The main driving point of this growth is the fact that the price of transportation of 1 ton of material decreases when the truck payload rises. At the beginning of the 1990s the payload became over 200 ton, and currently vehicles of 360-ton payload move in some pits. Some publications give consideration to the employment of 420-ton payload trucks (www.hutnyak.com). Some twenty years ago (Walker 1985) a forecast was made that at the end of the twenty-first century trucks could work at 900 ton capacity. Looking at the current state of development in this field, the time period for this prediction should be reduced by 50%. Nevertheless, some papers (e.g. Bozorgebrahimi *et al.* 2003, 2005) give indications that this development should slow down. Deciding factors that have an influence here are the increasing dimensions of transport roads and the mine as a whole, the fact that the greater sums involved need longer to recoup, the need for the enterprise as a whole to be large, and the greater system of technical indemnity needed (spare parts stores, maintenance shops, parking areas, fuel tanks, etc). Additionally, the progress in the construction of great wheels has met some significant drawbacks, ignoring here rapidly increasing tyre prices.

The durability of trucks of 80-ton payload and more is about 8–12 years or a little more; that is 200–300 thousand km (Church 1981). The mass of dump trucks is below its payload; for good constructional solutions, the ratio reaches 0.75. Dump truck prices are about 2.5 million US$ for units just above a 200-ton payload. For more than 300 ton, prices exceed 3 million US$. In large pits in the world, the number of applied trucks is about 100. The reserve size is usually about 15% of the total number (Hartman 1987). Codelco, Chile had 100 Komatsu dump trucks of 290-ton payload a few years ago. The largest system of this kind at present is in operation in El Cerrejón, Columbia where 17 power shovels and more than 150 trucks of 170-ton payload were working in the 1990s (Golosinski 1991). Some sources indicate that in a few years 28 hydraulic shovels will be in operation, working with 215 dump trucks (www.buengobierno.com/admin/files/ElCerrejon.pdf). It has been forecast that the system could be enlarged.

Taking into account the money involved in putting this kind of system into operation it is easy to conclude that it is a very expensive machinery system. Construction of an appropriately large maintenance bay costs a few million US$ and a fuelling system together with the necessary tanks and facility means that this expense almost doubles. Making an approximate assessment, it can be said that the fuel cost is about 30% of total expenses, while repairs consume a further 30%. The driver needs about 20% and the remaining expenses are connected with tyres (Church 1981). The prices of tyres are rising, and today the price of the largest wheels is over US$ 60,000. Their operation durability is in a wide range—from several weeks for tyres operating on bad roads, in tough conditions, and not properly cared for, to up to more than a year (Tattersall 2004). The use of used tyres also poses a problem.

The shovel-truck system[2] in widespread use in open pit mining needs appropriate technical support, not only in terms of stores, shops, tanks and in terms of a large amount of room for parking, but also additional machinery such as:

---

[2] The shovel-truck system is frequently referred to later on as the system or the machinery system.

- Water wagons to keep haulage roads in a good condition (in the dry season in tropical countries half an hour without water-spraying is enough to choke the whole system of trucks due to great amount of dust suspended in the air; truck wheels mill the road surface)
- Dozers for clean-up duties and supplementary actions in waste storage areas
- Front-end loaders for different purposes, also as spare loaders
- Towers, special transport vehicles to move failed trucks out from the pit
- Cranes, manipulators, etc. to perform maintenance, repairs, technical surveys and so on
- Equipment needed to maintain haul roads; this maintenance system consists of machines and appropriate devices
- Cisterns for fuel delivery.

At present, the majority of larger systems of this kind have dispatching centres. Previously, the main point of location of such centres was a suitable position at the edge of a hole in the ground with the necessary good view of the whole pit, and communication between the centre and drivers of trucks was by telephone. Nowadays, the centre is equipped with computers and sets of programs with data storage allowing analysis of the situation in pit, giving current and historical information on machines, their routes, time spent completing particular stages of the machine operation and many other things. Some computers also have simulation programs permitting some operation situations of the system to be generated; these are actual, historical and predicted scenarios. All hauling routes have sensors to trace each machine in motion on the road. Each machine can communicate directly with the centre. The truck dispatcher decides where a given truck should go or what to do, and where to go in the case of shovels. The decisions of dispatchers are supported by suitable information obtained from the computer. Generally, the system dispatcher possesses full information about where the given machine is, what it is doing now and what state the machine is in (repair, work, parking, etc.). In some cases, when changes of the extraction process should be modified, for example waste export, then overburden shovels might get priority. For acceleration of waste removal from the pit, the truck dispatcher takes strict care to keep all waste loading shovels continuously busy. After a certain period of time this situation can be reversed to speed up the mineral being hauled. In some pits, the location of machines is supported by GPS or hybrid systems (Russell 2006).

Truck dispatching relies on the current control of the system.

Three modifications of the above machinery system should be noted.

Several decades ago, the truck-inclined hoist system enjoyed some popularity. The number of trucks in such a system is reduced due to the application of hoists, although a certain number of haulers are still in operation as in the classical system. They move out material extracted from the pit. Some haulers deliver their load to measure the pocket of the hoist installation, where it is removed from the pit by a skip. The hoist is located on the part of the slope of the pit that will not be excavated; this slope will only be going down—extended. Due to significant development of truck technology, this system is rarely used today.

The second solution is transportation of broken rock to an in-pit crusher where lump sizes are reduced to a dimension that can be transported by a belt conveyor. Generally, the belt is the most expensive part of the conveyor as well as the most sensitive to the disadvantageous properties of the material being transported. In some cases a system with a crusher and belt conveyors is more profitable than the classical system with wheel haulers only (see Bingham Canyon in the late 1980s; Kammerer 1988). The number of trucks is reduced compared to the system with haulers only, but the cost of crushing can sometimes be high. Belt conveyors applied in this system can be conventional ones or HAC, sandwich-type for instance, if needed.

The third possibility in this field is the application in a shovel-truck system of a new type hoist—TruckLift. This installation can transport a fully loaded truck from the pit up to 480 ton of total mass. A truck travels, say, half an hour to get out of the pit; a hoist can make it in about 2 minutes. Nevertheless, a hoist can take only a certain part of the stream of haulers driving out of the mine. The greater the truck system is, the smaller the stream that can be taken. Moreover, application of the TruckLift system makes sense if the pit life is long and the whole enterprise appropriately large. Czaplicki (2004/2005) presented the procedure of calculation of this system: shovel-truck-inclined hoist of TruckLift type.

1.3 DESCRIPTION OF OPERATION OF THE MACHINERY SYSTEM

In this chapter, a short description of the operation of the machinery system discussed will be given in a verbal form. The reason for this is to prepare the reader for the difference between the real characteristic features of machinery operation in mines and the features of models applied by researchers in their publications. Comprehensive information on the literature published around the world over the last fifty years in this field is given in chapter 3.

The mining operation occurs in the pit. Overburden is constantly being removed and the surrounding rock is also removed to the necessary extent. This broken rock is waste. It is hauled out of the pit and located near the workings, waiting for the end of mining exploitation. It will then be used to fill the abandoned hole. The rock to be extracted is attained by blasting and the portions of broken rock scattered in the pit waiting to be loaded and hauled.

Loading machines—power shovels move from face to face to load. The time of the shovel work cycle, i.e. lowering bucket—filling it up—raising up the bucket—turning the machine to get the bucket just above the truck box—opening the flap/gate and emptying the bucket—return of bucket to broken rock by turning back the machine and shutting flap, is a random variable of symmetric probability distribution or not fully symmetric with a small positive asymmetry. This distribution has a mode. A few buckets fill the truck box. Time spent loading the truck is a random variable of similar properties. During the shovel operation, auxiliary works are sometimes necessary which shorten the time that the shovel can spend loading. These supplementary operations are usually divided (e.g. Church 1981, Czaplicki 2004—academic textbook) into short ones and long ones. These longer-lasting operations are such works as shovel displacement from one loading point to another, machine removal from the face due to blasting, etc. The short operations usually last a few minutes and are: verification of floor of loading area, loading of oversized boulder, etc.

The truck work cycle is a four-stage one as far as functioning is concerned, i.e. loading—hauling—unloading—return to loading machine. The times of these four random variables create probability distributions symmetric or almost symmetric with one mode. Trucks also operate in a shorter time than the whole shift. The time spent not hauling is connected with fuelling, change of drivers, coffee time and so on. Repairs are considered separately.

The whole system is controlled from the dispatching room, the main goal of this control being—generally speaking—to get the maximum profit from operation of the system, i.e. high utilization of machines involved and reduction of losses in operation time. For many years the main academic and research centres of the world have been conducting intensive research and proposed enhanced methods as well as new methods of truck dispatching—Lizotte and Bonates 1986, Soumis *et al.* 1986, Hagenbuch 1987, Wright 1988, Bonates 1992, 1993, Oraee and Asi 2004, for instance. A new light on truck dispatching is given in this book.

It has been assessed (Woodrow 1992, Bozorgebrahimi *et al.* 2003) that the cost of loading and hauling is approximately 60–65% of the total operating costs of open-pit mining, whereas haulage cost is about 40–50% (Wang and Zhao 1997, Woodrow 1992). Some researchers are of the opinion that the latter cost is even 60% (Fabian 1989).

CHAPTER 2

Queuing systems applied

In this monograph, modelling is the basic tool of theoretical analysis, which considers the reliability of both the equipment involved and the system created by these machines as well as the reality of a system operation process. Using proper modelling it is possible to discover the essential system characteristics and parameters as well as special local parameters that give vital information in certain performance areas.

A **model** is a pattern, plan, representation or description designed to show the structure or workings of a real or conceived object. In this monograph, consideration is given to the mine machinery system, with its specified stochastic properties and also its operation process when taking into account mine conditions. Observing this organized set of machines as a system it can be seen that it is dynamic, closed, cyclic, and under current control. The appropriate mathematical tools to describe such a system should therefore be investigated in the theory of queues.

Two models taken from this theory were used during the modelling and calculation of the system: the Maryanovitch model and the *G/G/k/r* model. The first model is described by Maryanovitch (1961; see also Kopocinski 1973) and originally involved a system of $n + m + r$ machines, where m machines were directed to work, n machines were hot reserves and r machines created cold spares. Because the hot reserve model is not needed in this study, it can be assumed that $n = 0$, and all patterns concerning this model are presented assuming the lack of hot spares.

The second model is a fundamental one; it has, however, been modified several times for adjustments of operation requirements. Generally, it describes the flow of machines in the system from a reliability point of view. This model is employed to describe the functioning of the truck system. Modification of the model is also presented.

2.1 THE MARYANOVITCH MODEL

A given system consists of m homogeneous machines directed to work and r machines in cold reserve (i.e. the intensity of failures of machines is negligible). It is assumed that working machines can fail with the constant intensity ε. A workshop in the system contains an adequate number of repair stands so that all failed machines can be repaired simultaneously (i.e. the number of repair stands is $m + r$). Therefore, there is no queue of machines waiting for repair. It is also assumed that repair times are independent random variables characterized by a general probability distribution $G(x)$. The expected value of this variable is known, and equals T_n. Additionally, three classical assumptions for queue systems are fulfilled (see for instance: Kopocinski 1973, Czaplicki 2004), namely:

a. Machines are served in a *FIFO* (first in, first out) regime
b. Repair entirely restores a machine's ability to work, and
c. Random variables of the model—work and repair times—are independent of each other.

A graphical representation of the Maryanovitch model is shown in Figure 2.1.

Some notations in Figure 2.1 need explanation.

The two transition intensities between states are designated by ε and η. The first one, ε, is the machine failure intensity in the system. Because it is constant, work times are characterized by an exponential distribution. Moreover, $\varepsilon = T_p^{-1}$, which means that this intensity is the reciprocal of the expected value of work times, T_p. The second one, η, is also the reciprocal, namely $\eta = T_n^{-1}$, but here there is an unspecified, general distribution $G(x)$.

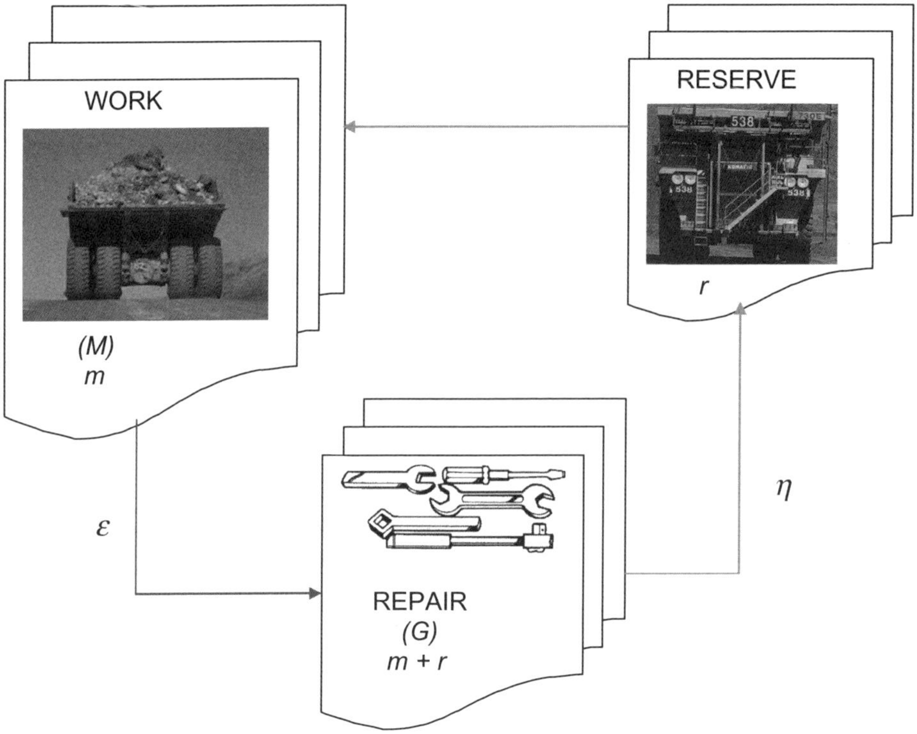

Figure 2.1. Operating scheme of the Maryanovitch system.

Two additional notations are shown in the figure (*M*) and (*G*), originally introduced by Kendall (1953). Kendall's notation, as it is known, is widely used in queues theory (see for instance Gross and Harris 1974, Kopocinski 1973, Sivazlian and Wang 1988, 1989, Czaplicki 2004). The symbol *M* denotes the exponential distribution (named after Markov) and *G* denotes any distribution (general case).

According to Kendall's notation, the Maryanovitch model can be given as:

$$M/G/m+r/r$$

This means that the arrival process to the service station is the Poisson one, the service time is of general distribution, the number of service stands is $m + r$ and the size of the waiting room is r.

It has been proved (see for example Maryanovitch 1961, Kopocinski 1973) that the system stated above has the probability distribution of a number of failed machines (i.e. machines in repair state) described by the formula:

$$P_k^{(n)} = \begin{cases} P_0^{(n)}(\kappa^k/k!)m^k & \text{for } k = 1,2,\ldots,r \\ P_0^{(n)}(\kappa^k/k!)m^r m(m-1)\ldots(m-k+r+1) & \text{for } k = r+1,\ldots,m+r \end{cases} \tag{2.1}$$

where: $\kappa = \varepsilon T_n$ is the failure rate, sometimes called the fault coefficient (Ryabinin 1976) and obviously

$$\sum_{k=0}^{m+r} P_k^{(n)} = 1.$$

The latter equation allows the set of equations (2.1) to be solved and interesting probabilities to be found.

This model has three important characteristic features. The first two are general, while the third concerns the mine situation. The first one is advantageous, the second disadvantageous and the third vital for mining purposes.

There is no queue of machines in the system waiting for repair. This is a very convenient property of systems of this type. In the practice of so-called mass servicing systems,[1] many systems need to be organized in such a way as to obtain this property whereby the queue of clients waiting for service is short or negligible.

The second property is the relatively large size of the service subsystem. If the number of machines in the system is not great, construction of a service station with a number of repair stands equal to the total number of machines in the system would appear rational. However, if the number of machines is big, as in large machinery systems of great open pits, say 80 or more than 100 units, construction of such a great maintenance bay makes no economic sense. Immediately, the question arises: if such a great service system is uneconomic, what is a convenient size of the system? This question is answered in the following considerations.

The third property is crucial for mine truck systems. As empirical reliability investigations have proved (Czaplicki 1989a, Temeng 1988, Czaplicki and Temeng 1989, for instance) work times of trucks can be satisfactory described by exponential distribution. The repair times have no such property, meaning that Palm's (1947, Kopocinski 1973) model cannot be employed, but only Maryanovitch's can.

2.2 THE *G/G/k/r* MODEL

This model is usually known as the Sivazlian and Wang model, after theories espoused by these authors (1988, 1989). It is a more general model than the classical repairman problem[2] (Kopocinski 1973) because it assumes that machines in reserve can fail.

There is one limitation in this model. The exploitation situation in the system must fulfil the so-called *heavy traffic situation*. This permits the description of the system by the method of diffusion approximation that is based on the assumption that queues of failed machines in the repair shop are almost always non-empty. If this condition is fulfilled, the discrete type of queue process can be replaced by the continuous type process. This change must be done in such a way that the characteristics of the original process will not be lost (Gross and Harris 1974). In contrast to the discrete space, the functions written continuously may be modified to give interesting characteristics. The result obtained is then transferred to the discrete space and the solution can be presented in explicit form.

Many of the exact solutions to queuing problems with inter-arrival times or service times distribution of the general type have not been found. It is extremely difficult to obtain explicit patterns such as the steady-state probability mass function, and the mean of the number of clients in the system for a *G/G/k/r* queuing system.

The first papers concerning the application of the diffusion approximation to queue systems can be dated to 1965 (Iglehart, Kingman). Iglehart considered the problem of limiting diffusion approximation for many server queues and Kingman reflected on similar issues. Glinski *et al.* (1969) discussed the application of the diffusion method for reliability forecasting. Heyman (1975) gave the diffusion approximation to the *G/G/*1 system. Halachmi and Franta (1978) developed a diffusion approximation model for the *G/G/R* queue that is consistent with some known

---

[1] This theory, born at the beginning of the twentieth century, was termed as the theory of queues or waiting line theory. In Central and Eastern Europe this theory has been termed as mass servicing theory. The latter term seems more appropriate.

[2] The Maryanovitch model is used in the particular case of the repairman problem too.

heavy traffic limit theorem. Yao (1985) applied the diffusion approximation model for the *M/G/m* queue. Kobayashi (1974) gave an application to the queuing network. In 1985 Haryono and Sivazlian carried out an analysis of a repair problem employing the diffusion approximation, and two papers published later by Sivazlian and Wang (1988, 1989) concerning the *G/G/k/r* system were comprehensive and significant not only for mine machinery systems. To prove this statement, it is necessary to consider their original model now. Later, some necessary modifications will be made to take into account mine reality.

A system of $m + r$ homogeneous machines is given. As many as m units operate simultaneously in parallel. The reserve size of these machines is r, but these spares are warm-standby units.

A standby component is called a *warm standby* or a *lightly loaded* if its intensity of failures α is non-zero but less than the intensity of failures of an operating unit δ, $0 < \alpha < \delta$. For $\alpha = 0$ the reserve becomes cold (unloaded) and for $\alpha = \delta$ it becomes hot (fully loaded). The repair is done with the intensity γ.

It is assumed that the work times probability distribution is the general one (G), the probability distribution of times to machine failure in reserve is general (G), and the repair times probability distribution too is general (G). The intensities of transition between states α, γ and δ are known.

Furthermore, the three classical assumptions of queue systems listed previously remain valid.

Figure 2.2. shows a graphical representation of the *G/G/k/r* model.

Considering the application of the diffusion approximation to find the mass probability function of the number of failed machines, Sivazlian and Wang concluded that three pairs of statistics

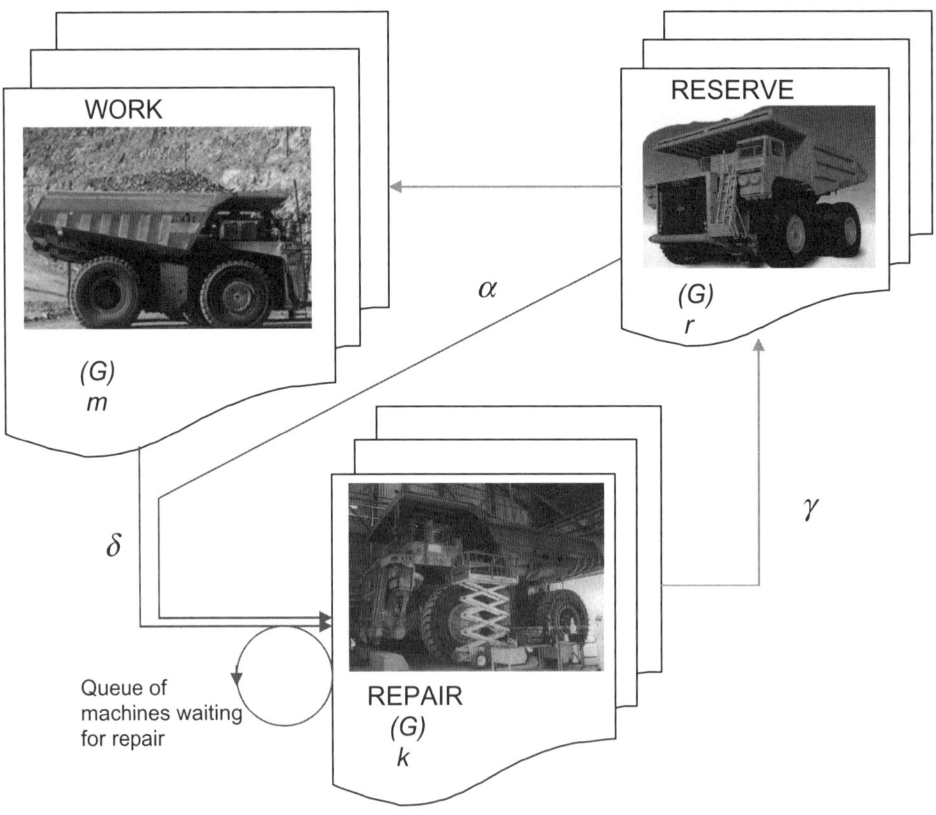

Figure 2.2. Operating scheme of the *G/G/k/r* system.

parameters are necessary. Apart from the named intensities α, γ and δ, the following standard deviations should be known:

- σ_p the standard deviation of work times
- σ_S the standard deviation of times that the machine spent in reserve up to the moment when failure occurred
- σ_n the standard deviation of repair times.

After transformation from the discrete space to the continuous one, the function that needs to be found is the probability density function $h(x)$ of the number of failed machines.

Using the following notations:

$$C_M = (\delta\sigma_p)^2 \qquad C_S = (\alpha\sigma_S)^2 \qquad C_R = (\gamma\sigma_n)^2 \tag{2.2}$$

and

$$\xi = \delta/\gamma \qquad \theta = \alpha/\gamma. \tag{2.3}$$

The parameters C_M, C_S and C_R are the square coefficients of a variation of the succession of the uptimes of the working machines, the uptimes of the spare machines, and the repair times, respectively.

The parameters ξ and θ are the failure rates for machines in the work state and in the reserve.

The probability density function of the number of failed machines x in the system expressed as the continuous function $h(x)$ is the sum of three components. They are defined as follows:

- For the reserve not lower than the number of repair stands, $k < r$

for $0 \le x < k$

$$h_1(x) = \frac{K_1}{m\xi C_M + (r-x)\theta C_S + xC_R}\left(\frac{m\xi C_M + (r-x)\theta C_S + xC_R}{m\xi C_M + r\theta C_S}\right)^{\beta_1} \exp\left(\frac{-2(\theta+1)x}{C_R - \theta C_S}\right)$$

for $C_R - \theta C_S \neq 0$,

$$h_1(x) = \frac{K_1}{m\xi C_M + r\theta C_S}\exp\left(\frac{2(m\xi + r\theta)x}{m\xi C_M + r\theta C_S} - \frac{(\theta+1)x^2}{m\xi C_R + r\theta C_S}\right) \quad \text{for } C_R - \theta C_S = 0 \tag{2.4a}$$

for $k \le x \le r$

$$h_2(x) = \frac{K_2}{m\xi C_M + (r-x)\theta C_S + kC_R}\left(\frac{m\xi C_M + (r-x)\theta C_S + kC_R}{m\xi C_M + (r-k)\theta C_S + kC_R}\right)^{\beta_2} \exp\left(\frac{2(x-k)}{C_S}\right) \quad \text{for } C_S > 0$$

$$h_2(x) = \frac{K_2}{m\xi C_M + kC_R}\exp\left(\frac{2(m\xi - k)(x-k)}{m\xi C_M + kC_R}\right) \quad \text{for } C_S = 0 \tag{2.4b}$$

for $r \le x \le m + r$

$$h_3(x) = \frac{K_3}{(m+r-x)\xi C_M + kC_R}\left(\frac{(m+r-x)\xi C_M + kC_R}{m\xi C_M + kC_R}\right)^{\beta_3} \exp\left(2(x-r)/C_M\right) \quad \text{for } C_M > 0$$

$$h_3(x)=\frac{K_3}{kC_R}\exp\left(\frac{2(m\xi+r\xi-k)(x-r)}{kC_R}-\frac{\xi(x^2-r^2)}{kC_R}\right)\quad \text{for } C_M=0 \tag{2.4c}$$

where:

$$\beta_1=\frac{2m\xi\left[C_R-\theta C_S+(\theta+1)C_M\right]+2r\theta\left[C_R-\theta C_S+(\theta+1)C_S\right]}{(C_R-\theta C_S)^2}$$

$$\beta_2=\frac{2m\xi[1-(C_M/C_S)]-2k[1+(C_R/C_S)]}{\theta C_S}\qquad \beta_3=\frac{2k[1+(C_R/C_M)]}{\xi C_M} \tag{2.4d}$$

- For a reserve greater than the number of repair stands, $r<k$

for $0\le x<r$ $\quad h_4(x)=h_1(x)$ (2.5a)

for $r\le x\le k$

$$h_5(x)=\frac{K_5}{(m+r-x)\xi C_M+xC_R}\left(\frac{(m+r-x)\xi C_M+xC_R}{m\xi C_M+rC_R}\right)^{\beta_5}\exp\left(\frac{2(\xi+1)(x-r)}{\xi C_M-C_R}\right)\quad \text{for } \kappa C_M-C_R\neq 0$$

$$h_5(x)=\frac{K_5}{(m+r)\xi C_M}\exp\left(\frac{2(x-r)}{C_M}-\frac{(\xi+1)(x^2-r^2)}{(m+r)\xi C_M}\right)\quad \text{for } \kappa C_M-C_R=0$$

for $k\le x\le m+r$ (2.5b)

$$h_6(x)=\frac{K_6}{(m+r-x)\xi C_M+kC_R}\left(\frac{(m+r-x)\xi C_M+kC_R}{(m+r-k)\xi C_M+kC_R}\right)^{\beta_3}\exp\left(2(x-k)/C_M\right)\quad \text{for } C_M>0$$

$$h_6(x)=\frac{K_6}{kC_R}\exp\left(\frac{2(m\xi+r\xi-k)(x-k)}{kC_R}-\frac{\xi(x^2-k^2)}{kC_R}\right)\quad \text{for } C_M=0 \tag{2.5c}$$

where:

$$\beta_5=\frac{2(m+r)\xi(C_R+C_M)}{(C_R-\kappa C_M)^2}. \tag{2.5d}$$

The probability density function $h(x)$ of the number of failed machines is determined by the formula:

$$h(x)=\begin{cases} h_1(x)+h_2(x)+h_3(x)=K_1g_1(x)+K_2g_2(x)+K_3g_3(x) & \text{for } k\le r \quad (2.6a)\\ h_4(x)+h_5(x)+h_6(x)=K_4g_4(x)+K_5g_5(x)+K_6g_6(x) & \text{for } k>r. \quad (2.6b)\end{cases}$$

In the construction of the above functions are the unknown constants K_i, $i = 1, 2, \ldots, 6$. Therefore, it is necessary to build 6 equations that allow these constants to be determined.

Function $h(x)$ must fulfil two conditions.

The first one is the normalization of the probability mass to unity.

Thus:

$$\int_0^{m+r} h(x)dx = K_1 \int_0^k g_1(x)dx + K_2 \int_k^r g_2(x)dx + K_3 \int_r^{m+r} g_3(x)dx = 1$$

and

$$\int_0^{m+r} h(x)dx = K_4 \int_0^r g_4(x)dx + K_5 \int_r^k g_5(x)dx + K_6 \int_k^{m+r} g_6(x)dx = 1. \tag{2.7}$$

The second condition assumes the continuity of $f(x)$. The following equations then hold:

$$K_1\, g_1(k) = K_2\, g_2(k) \quad \text{and} \quad K_2\, g_2(r) = K_3\, g_3(r)$$

as well as

$$K_4\, g_4(r) = K_5\, g_5(r) \quad \text{and} \quad K_5\, g_5(k) = K_6\, g_6(k). \tag{2.8}$$

Before these functions are converted into discrete space (the number of failed machines is the natural number plus zero) it is worth considering two cases.

The first case is when $C_M = 0$. This case was the result during the mathematical analysis. In engineering language, it means that the appropriate standard deviation of the work times of machines is zero. This does not hold in engineering practice. Therefore, in further considerations this case will be excluded.

The second case is the assumption that machines in reserve can fail. Fortunately, this is a rare case. Even if such cases may sometimes occur, the value of the corresponding intensity of failures is very small. It can therefore be ignored, and for that reason it is assumed further that $\alpha = 0$.

To return to the procedure, the corresponding function is obtained in the discrete space. There are several different methods of discretization. Here is the procedure proposed by Sivazlian and Wang following Halachmi and Franta's (1978) suggestion. According to the procedure cited, the steady-state probabilities $\boldsymbol{P}_j$, $j = 0, 1, \ldots, m + r$ are given by:

$$p_j = \int_{j-0.5}^{j+0.5} h(x)dx, \quad \text{for} \quad j = 1, 2, \ldots, m + r - 1,$$

$$P_0 = \int_0^{0.5} h(x)dx \quad \text{and} \quad P_{m+r} = \int_{m+r-0.5}^{m+r} h(x)dx \tag{2.9}$$

where P_j is the probability that j machines failed.

However, using these patterns it is necessary to pay special attention to the limits of determination of particular functions $h(x)$.

To close the discussion, a final problem needs to be considered. This is the criterion that should be fulfilled to validate the system's operation under the heavy traffic condition.

The first idea is that the probability that the repair shop is empty should be small. This condition is inconvenient. The service station possesses k stands and the heavy traffic situation requires occupancy of the majority of these units almost continuously. Thus this condition can be expressed by means of system parameters. According to Halachmi and Franta (1978) the decisive factor is the fulfilment of the following inequality:

$$\rho = \frac{\Lambda}{k\gamma} \geq 0.75, \tag{2.10}$$

where Λ is the intensity of arrivals to the service station and ρ is the coefficient of flow intensity in the station (Gross and Harris 1974). It is worth noting here that the above condition has no strict meaning. It has been stated that if the coefficient tends to fall below 0.75 the assessment of probabilities (2.9) becomes poorer. It is also worth noting that the coefficient of flow intensity in the service station should also fulfil the inequality:

$$\rho < 1, \tag{2.11}$$

which means that the intensity of arrivals to the service system should be less than the intensity of service in the system. Otherwise, an almost permanent queue will be observed, clients will be unsatisfied, and trouble will be inevitable.

CHAPTER 3

Literature review

The last fifty years have been characterized by intensive development in publications looking at non-operational methods of investigating the systems of machines and technical devices as well as installations applied in mining. This concerns analytical methods as well as simulation methods, but these cannot exist without good empirical data.

Probably the first paper on queuing theory applied to mining was a paper written by Koenigsberg (1958). The problem that he analyzed concerned the determination of production for a set number of crews working at the faces of several underground mines.

The 1960s saw serious progress in simulation techniques applied in mining problems; witness Rist (1961), Teicholz (1963), Bishele *et al.* (1964), Harvey (1964), Aurignac *et al.* (1968), Bucklen *et al.* (1968), Eichler (1968) and Juckett (1969). These papers were initially about mine machinery systems of continuous operation; later rail transport was included, and later still, vertical transportation was added. Finally, truck haulage in open pits was examined (Madge 1964, Cross and Williamson 1969). In investigations of conveyor systems, the time slicing method was employed, while the sequence of events method was used a little later. For rail transport and hoist transport, time slicing was used exclusively. A very good source of information in this regard is the proceedings from the APCOM—Application of Computer and Operations Research in Mining. The first event of this kind was held in 1961 at the University of Arizona, USA. It was also at the beginning of the 1960s that the first publications on the application of queue theory to cyclic machinery systems appeared. So-called Technical Reports were issued, such as those by Spaugh (1962), Teicholz and Douglas (1964) and Gaarslev (1969). Their considerations concerned two-link transport systems. One link was the loading or unloading point, and the second units moving between the first link and the point of their destination. The stochastic streams identified were of the Poisson type, as they were in the Koenigsberg paper. The publication by O'Shea *et al.* (1964) was based on the same assumption. Morgan and Peterson (1968) estimated shovel-truck system output, identifying the functioning of a truck as a four-stage process. The probability distributions were assumed to be exponential. They dismissed the application of the Palm model as too simplified. The Palm (1947) model is $M/M/k/r$ and a two-stage one. In Poland Kopocińska (1968) applied the Takács model (1962) for analysis of the shovel-truck system operating in a gravel mine. The model presented was a two-stage one, and it was identified as $M/G/1/m$ according to the Kendall notation. The problem of the reliability of machines did not exist in the above papers.

In the 1970s, many publications appeared in this area. Graff (1971) discussed the problem of the application of the queuing model with an unlimited source of arrivals and times of an exponential character. Connell's *SME Mining Engineering Handbook* (1973) presented comprehensive musings on the great merits of stochastic simulation. Huk and Łukaszewicz (1973) comprehensively discussed Kopocińska's problem, applying a digital simulation, but the service point was still alone. Kapliński (1974) considered the system $M/M/1 + m$ with a limited incoming stream to an analysis of the systems applied in civil engineering. Similar systems are employed in mining. Maher and Cabrera, in several few papers (1973, 1975), Tseng (1973) and Elbrond (1977) examined similar problems—a mixture of deterministic and probabilistic orders of events was taken into consideration, adding some modifications for particular cases. Barnes *et al.* (1979) expanded upon the considerations made by Morgan and Paterson. The service size was taken into account together with a comprehensive analysis of stages of truck movements. Barnes *et al.* applied the latest results in the theory of queues published by Gross and Harris (1974).

After the great success of the application of the theory of continuous machinery systems in mining supported by the simulation technique in Poland in the middle of the 1970s, Sajkiewicz, the co-author of this success, published in 1979 a textbook on the calculation of both continuous and cyclic machinery systems employed in mines. It should be stated that the part connected with cyclic systems was not well aimed—Markov processes were exclusively applied.

Developments in application and improvements in the simulation technique itself were observed in the same decade. Quite a number of works were published, such as those of Firganek and Wianecki (1970), Bauer and Calder (1973), Manula and Rivell (1974), Newhart (1977) and Talbot (1977).

The 1980s were years of great scientific output in this field. At least twenty publications are well worth mentioning. Generally, these papers can be divided into three categories. The first group contains papers dealing with the simulation technique. Quite a wide range of verified and improved models concerning the analysis of the continuous machinery systems and the first models relating to cyclic systems were the bases to develop enhanced simulation methods. Looking at these methods now, they can be assessed as relatively poor, but computer techniques at that time were at an early stage in their development. In Canada at the beginning of the decade, significant research was devoted to so-called interactive computer modelling for application in mining: Nenonen *et al.* (1981), Hufford *et al.* (1981), Chan (1982a, b, c) and Nenonen (1982). The probability distributions of phases of the truck cycle were identified. These are recognized as the normal, but Hufford *et al.* (1981) indicated that the lognormal distribution could also be applied. A similar opinion was given by Griffin (1989), considering the application of perturbation theory to the simulation of a truck-shovel system. In this paper he stated that the application of exponential probability distributions for description of times of truck cycle phases is *unrealistic*. During discussions held during the Mine Planning and Equipment Selection Symposium in Calgary, it was stated that the model in which the exponential probability distribution appears to describe the distribution of times of truck movement phases becomes useless. Lizotte *et al.* (1989) carried out a review of the applied methods of truck dispatching; the assessment was done by means of the simulation technique. They introduced the three-parameter Weibull probability distribution to describe times of truck loading and unloading. Szymanski and Srajer (1989) described the monitoring system supported by simulation technique for trucks using a SLAM II program.

The second category is made up of papers discussing the application of queue theory. In Poland, Stryszewski (1981) published a paper in which the parameters of the shovel-vehicle-crusher system were analyzed based on the exponential probability distributions of times on the random variables being examined. Barbaro and Rosenshine (1986) evaluated the productivity of a shovel-truck system using a cyclic queuing model. They corrected the wrongly formulated pattern of Barnes *et al.* and applied the simulation to verify the results obtained. The employed distributions they obtained were, however, exponential ones. They were aware of the fact that these distributions were not proper ones, but tried to show that the parameters obtained were not bad. How great an error was made was not stated. In 1987, Carmichael published a comprehensive elaboration on application engineering queues in construction and mining. The greatest advantage of the book is its presentation of several empirical machinery parameters and some relationships traced during field investigations. Data were mainly obtained from Australian mines. The key statistical tool was again exponential distribution. The deterministic order of events was also considered. However, some problems were discussed by decomposing stages on a certain number of phases, i.e. the Erlangian approach. The distributions of the Erlang type applied had the value of a shape parameter up to 50. The reliability problem of operating machines was briefly considered, but it was a separate problem not connected with queue models. In reliability considerations, the exponential distribution was again the main mathematical tool. The truck reserve size was considered as well but the relations obtained were connected with the employment of the exponential approach. It is worth remarking here that the author mentioned the possibility of a truck queue occurring before the repair shop. Zhongzhou and Qining (1988) proposed to apply the Erlang distribution to describe the times of truck work cycle phases. The model presented in this way was an $E/E/k/r$ type with a two-phase truck work cycle: loading and travel. The discussion comprised Erlang distributions up to the 6th order, assuming that these would be useful for practice. However, it is a well-known fact (see Gertzbach and Kordonsky 1966) that the transition from gamma distribution (of which Erlang distribution is a special case) to the normal distribution makes sense if the shape parameter (order) is not less than 9. From the theoretical point of view this approach was not new—Kopociński (1973) described the possibility of the application of Erlangian systems for engineering purposes. These systems have their advantages and disadvantages. These were described by Czaplicki (2004).

At the end of the 1980s some further publications appeared: Fabian (1989) looked at the application of queue systems for mining engineering purposes, but the distributions considered were exponential; and Czaplicki (1989) suggested considering the truck work cycle as a two-phase model with the probability distribution of times for one stage as exponential, and the second phase as the sum of three different exponential random variables. Around the same time Sharma (1989) and Sharma and Ekka's (1989) papers were published. These presented a different approach to the analysis of shovel-truck systems. They devoted their attention to analyzing the data obtained during field investigations. They constructed some operation measures of the system, indicated which measures are more important, and made an analysis of the system considering the *goodness* of changes in it. The main parameter was the system output. They included changes in the number of trucks applied. Kumar and Sinha (1989) described the operational reliability of haulage means for a particular mine.

Generally, in analytical models machines were completely reliable, while in simulation systems the reliability of trucks was often included. During field research, the reliability of the equipment under investigation was of course taken into consideration.

In the 1990s, the number of publications dropped in relation to the previous decade. Panagiotou (1993, 1994 and 1996) considered the application of simulation technique for analysis of a system. Kozioł and Uberman (1994) made an analysis of the system employing exponential distribution and Poisson-type process. Kumar (1996) described a maintenance strategy for mine systems. Frimpong *et al.* (1997) presented the functional model of the maintenance of mining equipment. Exceptions to the rule were Czaplicki's publications (1990, 1992a, b, 1992, 1994, 1997 and 1999). The main difference was the application of the Sivazlian and Wang model (1989) for modelling and analysis of the system. Sivazlian and Wang considered comprehensively the *G/G/R* machine repair problem with warm standbys employing the diffusion approximation. Czaplicki (1990c) analyzed the system: one loading shovel and a certain number of trucks. The main point of interest was the probability distribution of a number of trucks at the shovel. Complicated patterns were constructed. Two years later his considerations were orientated towards the probability distribution of the number of working trucks per one working shovel. This was probably the first paper in which reliability indices were included in an analysis of the system. It is worth remembering that one of the cardinal features of machinery systems is that the number of working machines at a given moment of time is a random variable. Taking into account the fact that the system has a truck reserve and different publications supply diverse recommendations, Czaplicki (1992b and 1994) considered the problem of truck spare units tracing basic regularities having an influence on the reserve size. The reliability of trucks was included in the discussion. The same author, in a paper from 1993, conducted an analysis of the operations of the system applied in quarries: front-end loaders and hauling trucks. The model employed was $M/G/m + r/r$, the correct one for small systems of this kind. Different possibilities for increasing the system output were examined. A more general case was considered in a paper dated 2000, assuming that the system is non-existent and that machines of unknown reliability have to be bought. It was assumed that the reliability parameters are random variables. Czaplicki (1997) analyzed carefully the system discussed by Morgan and Peterson, Barnes *et al.* and Barbaro and Rosenshine, applying different models for different sets of incoming parameters. The research comprised the deterministic model, Palm model, Takács model, Barnes *et al.* model, up to the *G/G/1* queuing model. Next, the assumption was made that the machines involved have a certain reliability and the results obtained were duly corrected. In conclusion, a practical recommendation was offered concerning the pair of numbers: the number of trucks directed to work and number of trucks in reserve. In 1999, Czaplicki published a paper in which a new method of shovel-truck system calculation was presented. All major components of the problem were included: the reliability of machines, their accessibility and the existence of a truck reserve. The proposed method was based on the probability distribution of a number of working trucks per one working shovel. However, further investigations proved that this approach gives *averaging* characteristics, of a different shape to the original ones. This method was generally corrected in the publication of 2002. Besides reliability and accessibility of machines and truck system structure, decisions made by the truck dispatcher were also taken into account. The method in question gives approximate solutions, if the heavy traffic situation is fulfilled. In large mine machinery systems it works. This paper was the culmination of many years of investigations. Further publications dated 2004 and 2005 enlarged on this topic. The first paper, which shows the method of analysis and calculation of the shovel-truck system expanded by a crusher and conveyors subsystem, is the only one considering this kind of problem in this way. Similarly, it presents the method of computation of a shovel-truck system additionally equipped with the inclined hoist of a TruckLift type. To date no publication in this field exists.

As for publications of the new century, at least the following are worth discussion.

A significant book was published in 2000, written by Sturgul. This publication was orientated to mine design applying the simulation technique. Awuah-Offei *et al.* (2003) discussed the application of the well-known program SIMAN for particular mining conditions. Kolonya *et al.* during the Mining Congress in 2003, presented a new method of shovel-truck system simulation taking into account the crusher application, whereas Nanda (2003) considered some calculation models employing queuing systems. The distributions discussed were exponential. In 2004, four papers were issued: Oraee and Asi, Bascetin, Kuruppu, and Weicheng and Youdi. The first one judged the possibility of the application of a fuzzy model for truck allocation in surface mines. The second work discussed the use of the analytic hierarchy process in equipment selection. The third paper described the problems of maximizing the reliability of equipment employed in mines. The last considered the opportunity of using the genetic algorithm to optimize the number and size of equipment for the machinery system of a surface mine. The authors of the first paper stated that the results obtained should be treated carefully, being just the basis for further analysis. The second paper presented a new approach to the discussion of mining problems. During the Mining Congress in Teheran, Ataee-pour *et al.* (2005) presented a new model of analysis of the system by means of a simulation technique to analyze the system employed in an iron-ore mine in Iran.

At the end of this literature review, it should be stressed that simulation methods are generally well developed, while analytical methods are far from the reality, with the exception of the works of Czaplicki. This is in spite of the fact that simulations are quite advanced techniques. However, taking into account all stochastic regularities and other rules deciding on the course of the exploitation process[1] of the system is still a hard thing to accomplish. Some necessary shortcuts are usually made, together with some simplifications. It can be seen just how many stochastic and deterministic phenomena must be taken into consideration after reading this monograph. Besides, simulation techniques have their own, sometimes unpleasant, disadvantages.

[1] The term 'exploitation process' will be defined in chapter 5.

CHAPTER 4

Purpose, method applied and field of consideration

In the theory of mine machinery systems there are four types of systems. They are distinguished by their method of functioning (Czaplicki 2004). They are:

- Continuous systems (systems of continuous technological structures)
- Readiness systems
- Cyclic systems
- Mixed systems.

The theory and analytical description of continuous systems was developed mainly in the 1960s and 1970s and comprises both formal models and simulation models. These were successfully verified by practice in mines, and so the mining world is in possession of a vast range of confirmed mathematical tools to describe and analyze systems of this type. Further progress in this field is observed in the development of new enhanced methods of simulation.

The development of the theory and analytical description of readiness systems is strongly connected with military problems. The majority of military systems are just readiness ones, i.e. systems that exist with the main purpose of expectation and readiness for action (fortunately). All rescue systems—fire brigades, police, medical emergency services, etc.—are examples of such systems. In mining, rescue systems in place to release underground miners trapped by a roof collapse are extremely important. Generally, the mathematical models applied are well developed and confirmed in practice. The main point of consideration for these systems is the problem of achievement of appropriate values of system parameters. This problem is a two-dimensional one. Readiness systems are usually systems of people and appropriate equipment. If people are well trained in applying their tools, the system attains suitable values of its parameters. On the other hand, a problem can be posed by the correct estimation of these parameters.

The theory and analytical description of mine cyclic systems are divided into four types:

- Operation of hoisting installations
- Operation of railway systems
- Operation of systems: loading machines—hauling units
- Operation of systems involved in rock extraction by blasting.

The mathematical tools applied to describe and analyze the operation of these systems are various.

Analytical description and analysis of the operation process of mine hoists are not the problem, apart from the description of cooperation between horizontal transports with a hoist. The existence of a shaft bin in such a system, while very convenient for mine practice, greatly complicates the analytical description. Apart from simulation methods, there are no effective analytical tools for a horizontal transport means—shaft bin—hoist system. Czaplicki (2005) performed a certain analytical trial of such a problem.

Analytical description and analysis of the operation process of mine railway systems are well known and verified, and give generally correct results. An improvement can be made by introducing in greater extend the stochastic phenomena existing in the operation process.

An example of a loader-hauler system is the shovel-truck system. Its analytical description and analysis have not previously been fully put into the context of mine practice. The great complexity of the phenomena featured in this system during operation as well as the difficulty of analyzing the stochastic properties of these phenomena have meant that almost all mathematical models were a weak approximation of reality. Some models which had the advantage of having a close relationship with mine practice, usually considered just a small part of this reality. This statement was proven in the literature review of the previous chapter. It is important to be clear that the role of this kind of machinery system in world mining today is significant. In the construction of open pits and the equipment operating there, larger and larger sums of money have been involved. The enterprises, which can be observed in this type of mining, are becoming more and more spectacular.

Therefore, taking into account the current state of world mining development, trends in the growth of large mine machinery as well as the considerable inconsistency between the possibilities of analytical description and analysis offered by theoretical models and the real properties of the system, this book presents the procedure used to model a system, taking into account the majority of stochastic phenomena existing in it, and to calculate the essential measures of system efficiency. This is the goal of this work.

The basic mathematical tool therefore is stochastic modelling, which relies on the construction of sequent analytical models, which, in determined succession, create the whole procedure. These models are based on the real properties of the operation process of system as well as on the real properties of machinery engaged. Thus, modelling is the describing and analyzing tool for the entire system and for its subsystems as well.

The scope of consideration of this monograph comprises the following problems:

1. Identification of the probability properties of components of the exploitation process of the system under consideration as well as the system structural elements (chapter 6)
2. Creation of a procedure, through construction of sequent mathematical models, allowing the characteristics of the system to be obtained. The modelling is divided into two parts; the first part comprises:
   - Construction of the reliability and functioning models for the shovel system (section 7.1)
   - Construction of functioning of the truck-workshop system, taking into account the reliability of trucks and service capacity of the workshop (section 7.2)
   - Construction of the probability distribution of the number of trucks in work state (section 7.3)—key information needed in later stages of modelling
3. Assessment of the *goodness* of selection of the applied machinery system matching the required transportation task—verification of the selection of the structural system parameters $< m, k, r >$, i.e. the number of trucks directed to accomplish the transportation task, the number of trucks in reserve and the number of repair stands in the workshop (chapter 8)
4. Creation of the second part of the modelling, considering the following problems:
   - Construction of the model, taking into account the reliability of repair stands (section 9.1)
   - Construction of the model operation of the shovel-truck system, taking into account the reliability of its machines, their accessibility and decisions made by the truck dispatcher (section 9.2)
5. Appearance of the system calculation, i.e. construction of a set of measures allowing an estimation of system efficiency in a wide sense; this is made together with a further analysis of the system (chapter 10)
6. Presentation of an example of how a particular system can be modelled and computed through comprehensive analysis (chapter 11)
7. Consideration of spare loading machines (chapter 12) and changes in the calculation procedure due to their application
8. Presentation of an example to illustrate the calculation procedure and analysis when spare loaders are applied (chapter 13)

   The whole modelling procedure is presented in graphical form in Figure 4.1.

9. Consideration concerning a selected problem in the truck-dispatching—the system with priority—and presentation of how the effects of this dispatching rule can be found in the analysis and calculation procedure (chapter 14)
10. Consideration directed towards the problem of how the system characteristics change due to changes in the hauling distance (chapter 15).

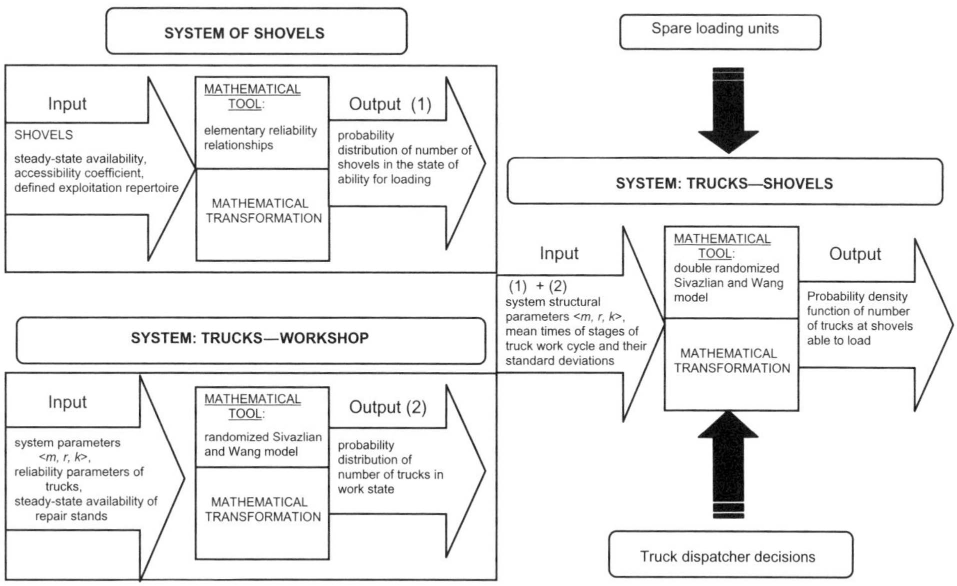

Figure 4.1. Scheme of the modelling procedure.

Two additional chapters are included in the monograph.

In chapter 5, the basic terms of reliability theory are identified and defined in order to make the whole considerations more communicative. This chapter also discusses the difference between theory of exploitation and terotechnology and why the expression *exploitation process* is frequently used in preference to the word *operation*.

Chapter 16 looks at a special topic—the availability of a technical object. This section is included because in many English engineering papers different definitions are given and different types of availability are considered. This chapter summarizes the existing situation.

The book is closed by the final remarks given in chapter 17, the list of all references cited, and the index.

CHAPTER 5

Reliability and the exploitation process

The subjects analyzed in this book are selected properties of mine machinery systems, properties partly of a stochastic nature. These are connected with the equipment involved and are expressed by reliability and operation parameters. Other partly stochastic properties are associated with the system operation (functioning) and are expressed by the operation parameters exclusively. It will therefore help if a few definitions of some important terms of reliability theory are recalled. Special attention will be given here to the theory of exploitation—terotechnology. This needs some explanation because diverse definitions are given in different publications.

Reliability** of an object (an item)* ***is *its* ***ability*** *(property, feature)* ***to fulfil requirements*** *(to maintain its functions, to accomplish its task)*; see for instance: ARINC Research Corporation (1964), Kiliński (1976), Migdalski ed. (1982), Wikipedia, DIN 40041, Kozlov and Ushakov (1966), Atis Telecom Glossary (2000) and Malada (2006).

This definition is the analytical, descriptive version. By including measures of this ability the definition changes to a normative one, i.e. the reliability of an object is its ability determined by values of significant parameters describing this feature to fulfil requirements.

Notice that an object applied in different conditions to fulfil different tasks usually has different values of reliability parameters (indices). For example, an armoured flight conveyor operating on a coalface and the same conveyor applied somewhere in a haulage line between two belt conveyors, operating as the breaking element because transport is going down and in order to avoid coal sliding down during transportation. These two identical conveyors will have significantly different reliability parameter values. Therefore, very often the definition is extended to something like the following:

'The ability of an object to perform a required function under given conditions for a given time interval' (BS3811, Federal Standard 1037 C).

Reliability is a property of a statistical nature, which means that the parameters stated above are taken from the theory of probability, e.g. mean work time to failure, probability of survival, etc.

Reliability theory divides all objects into two categories, taking into account two criteria of division:

a. Discrimination of object components
b. Property of restoration of object ability to fulfil requirements.

From criterion (a) there is a division of objects into:

- Elements, or
- Systems.

An element is an object that is undividable from the considered point of view.

A system is an organized set of elements.

Criterion (b) gives:

- Objects working until their first failure (irreparable)
- Objects that can be repaired.

Obviously, all the technical objects considered in this book can be repaired, meaning that the operation process of a particular object will always be at least two-state: work—repair.

From the point of view of reliability, the operation process of each machine considered here has two types of events:

- Failures, and
- Renewals.

and two types of random variables:

- Work time between two neighbouring failures (for short: work time), and
- Repair[1] time.

A graphical description of such a process with these events and random variables is shown in Figure 5.1.

Reliability theory assumes that the properties of an object, determined originally on the design stage of the object and finally on its production stage, are one of three main factors[2] deciding on the course of the operation processes of an object. During the work state the instantaneous potential (defined by these properties) to fulfil requirements of this object is exhausted and, finally, failure occurs. During repair this potential is restored. Repair is finished and the work resumes. These two states—work and repair—are called *the own states* of an object. They depend strongly on the properties of the object.

Two basic reliability parameters are needed to model and calculate a shovel system, namely the expected value of each random variable stated above. Recall that the expected value E of a random variable, say $\boldsymbol{X}$, is defined as:

$$E(\boldsymbol{X}) = \int_x xf(x)dx. \tag{5.1}$$

These parameters are also required to calculate the shovel steady-state availability.

Four basic reliability parameters are essential for every truck considered during modelling and calculation of the system. These are:

- Expected value of both random variables
- Standard deviation of both random variables.

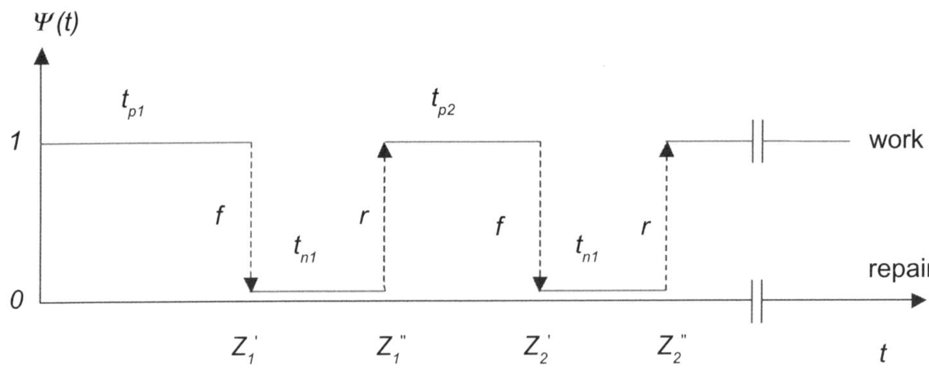

Figure 5.1. Process $\Psi(t)$ of changes of states: work—repair.
Keywords: f—failure, r—renewal, t_p—work time t_n—repair time.

[1] The term 'repair' state is commonly used in literature. In BS3811 'fault' state is used, but this is rare compared to repair.
[2] Two other factors are: applied methods of object utilization and maintenance, and the environment of the object.

Recall that the standard deviation $\sigma(X)$—being the measure of dispersion of values of random variable around their mean—is defined as the positive square root from the variance $\sigma^2(X)$ that is defined by the equation:

$$\sigma^2(X) = E[X - E(X)]^2. \quad (5.2)$$

At all stages of modelling and calculation a very important reliability parameter is needed, sometimes named 'availability' for short. Because of the existing state of mining engineering literature, a special topic is discussed in chapter 16 of this paper. So, it is assumed—for the time being—that this term is clear and measures of availability are known.

The second field of science—besides reliability theory—that will be widely applied in this consideration is the **theory of exploitation**. This expression is well known in Central and Eastern Europe. In the English-speaking world, one can use the interchangeable term **terotechnology**. Both terms mean roughly the same, but it is hard to join them together because of the difference in how a particular problem should be approached as well as what the scope of consideration and main points of interest are.

The history of early development was different in the United Kingdom, where terotechnology was born, and in Central Europe. However, the period of delivery is roughly the same—the end of the 1960s and beginning of the 1970s. In the United Kingdom, the problem of maintenance of objects separated partly from reliability when national conferences were held on this topic in the 1960s. In 1967 the British Council of Maintenance Associations was born. After several years of vast development of works of both an empirical and a theoretical nature, the Committee for Terotechnology was formed. It took two years to produce a definition of terotechnology. 'This is a combination of management, financial, engineering, and other practices applied to physical assets in pursuit of economic life cycle costs. The practice of terotechnology is concerned with the specification and design for reliability and maintainability of plant, machinery equipment, building and structure with their installation, commissioning, maintenance, modification and replacement, and with feedback of information, performance and cost' (Hewgill and Parkes 1979).

'In the future terotechnology will be an essential element of good husbandry, of quality and of the ability to understand that an artefact commits resources both in its making and in its subsequent use ... the outcome of such an approach may result in a product which has high initial cost and long reliable life, or which is cheap with a short life and anticipated replacement or breakdown ... Terotechnology has a simple objective—that of minimizing the whole life cost of ownership—but its practice can be complex, involving interdependencies and relationships of a diversity of resources—people, money, material, ideas and techniques' (Darnell 1979).

BS3811 was published in 1993.

For an overview of terotechnology today, three quotations should be considered.

The British standard cited above gives the following definition: 'a combination of management, financial, engineering, building and other practices applied to physical assets in pursuit of economic life cycle cost'.

Bhaudury and Basu's (2002/2003) book on terotechnology—probably the first in English—stated succinctly: 'Terotechnology—a concept, nay, a philosophy.'

According to the online MSN dictionary (2007) it is: 'a branch of technology that uses managerial and financial expertise as well as engineering skills when installing and running machinery.'

Investopedia (2007) says that it is: 'A word derived from the Greek root word "tero", or "I care", that is now used with the term "technology" to refer to the study of the costs associated with an asset throughout its life cycle—from acquisition to disposal. The goals of this approach are to reduce the different costs incurred at the various stages of the asset's life and to derive methods that will help extend the asset's life span. Terotechnology uses tools such as net present value, internal rate of return and discounted cash flow in an attempt to minimize the costs associated with the asset in the future. These costs can include engineering, maintenance, and wages payable to operate the equipment, operating costs and even disposal costs'.

Similar ideas are presented by Belak (2005).

In Central Europe, mainly in Poland, around the end of the 1960s and the beginning of the 1970s, after the spectacular development and popularity of reliability theory, researchers quickly concluded that many reliability problems associated with the use of an object in real conditions need to be considered within a wider scope. It was necessary to consider some properties of the object not connected with reliability in a direct way but connected with object properties associated with the process of its application in particular operation conditions. Researchers were interested in what was going on with many object properties during the use of the object, and later, during maintenance. In this way, the theory of exploitation came into being. A great part of reliability considerations belongs, therefore, to the scope of consideration of the theory of exploitation, but the scope of considerations of the theory of exploitation is much wider than that of reliability.

The term exploitation comes from the French word *exploitation*, which means *usage*. In connection with engineering problems, exploitation is based on the statement that it is the *usage of something in a rational way*. But in the English-speaking world, the term *exploitation* possesses many negative connotations. Looking at Thesaurus.com (2007) one finds synonyms such as: 'dishonesty', 'crime', 'misuse', 'cheat', etc., and even 'unwanted sexual advance', and Wikipedia (2007) gives an association with Marxist theory. For this reason acceptance of the term *exploitation* is extremely difficult in this part of the world.

In the mid-1970s, much was published on the topic in Central Europe, and after several years of intensive development, this theory was well stabilized. Researchers involved in this kind of investigations came to the conclusion that 'the exploitation of a technical object is a set of intentional actions of a technical, economical and organizational nature directed at this object, as well as mutual relationships existing between them (people cooperating and the object) from the moment of the object's first usage until its withdrawal and disposal' (Polish Standard PN-82/N-04001). Similar definitions can be found in the standards of neighbouring countries. For more than twenty years, in a number of technical universities the theory of exploitation has been offered as a separate lecture subject in connection with the utilization of various technical objects. Many books and textbooks have been issued (e.g. Adamkiewicz 1982, Downarowicz 1997, Kaźmierczak 2000, and Będkowski and Dąbrowski 2006). The theory of exploitation[3] has proved very useful.

The general problem of exploitation theory is:

What to do or how to arrange the path of the exploitation process of an object in order to obtain its most convenient course?

Other points of interest are the answers to many particular questions like:

- *What are the components of the exploitation process? How can they be identified?*
- *What kind of influence does the object environment have on the course of the exploitation process of it?*
- *What are the technical and economical possibilities for changing the course of the process?*
- *What kind of changes will be the most profitable?*
- *What kind of changes in the construction of the object or system structure should be carried out to improve its achievements?*
- *How should the course of the process be arranged to assure an appropriate level of safety?*

and many, many others.

The main point of interest here is a certain course of action called the *exploitation process*. This is a process of changes of the properties of an object during its utilization and maintenance. In the English-speaking world, the term *operation* can also be used. However, this word appears to be over-used. Thesaurus.com (2007) gives 48 meanings of this expression; i.e. almost 48 definitions should exist for it. It seems it is a picklock word. In science, precise terms[4] are needed.

[3] An exploitation process consists of two processes: utilization and maintenance, interlacing with each other.

[4] BS3811 gives the following definition of the term operation: 'the combination of all technical and administrative actions intended to enable an item to perform a required function, recognizing necessary adaptation to changes in external condition'.

Here is a short summary.

Terotechnology is a *concept, combination of practices, kind of technology, philosophy*.

Terotechnology has the objective:

to minimize the whole cost of ownership of the object life.

Theory of exploitation is the strictly defined *field of science* determining the fundamentals of exploitation of objects in a rational way.

Theory of exploitation has the objective of comprehending the mechanisms governing the course of the changes of object properties during its utilization and maintenance. It is also interested in minimizing the whole object life's cost of ownership, if the safety requirements are fulfilled.

Therefore, in the later calculations, the theory of exploitation will be applied, but the term operation will be used relatively frequently to make the text more understandable for English native speakers.

To return to the discussion.

The investigation of properties of the exploitation process of the machinery system under consideration sometimes concerns properties that can be identified immediately with regard to measure; it is enough to observe symptoms occurring during the process of exploitation. However, there are several further properties that can be identified and measures that can be attributed only when handling appropriate data. Sometimes modelling is very useful here, because working with the model allows interesting measures of the process to be constructed. The term exploitation process will be understood here in a slightly wider sense than that usually identified in the theory of exploitation. According to the classical definition, the exploitation process of an object is the process of changes of its properties from the moment when utilization is commenced until the moment of definite withdrawal of the object utilization, i.e. during the object life cycle. Kaźmierczak (2000) (p. 156) formulated a definition of an exploitation process that is ideally suited to this discussion: 'an exploitation process is everything that happens with the object from the moment of the end of its production till the moment of its withdrawal from utilization.'

The exploitation process of the shovel-truck machinery system being examined will be understood as a two-dimensional process:

- Process of changes of reliability properties of machines that the system consists of
- Process of changes of functions executed by machines.

These unit processes are not equivalent to each other.

The course of the process of changes in reliability properties does arise from the process of functioning, but not exclusively. Two other factors have an influence here—exploitation conditions and object properties given during stages of its design and production. The process of changes of reliability properties is superior to the process of functioning.

There are two basic terms of exploitation theory associated with the term 'exploitation process'. These are: *state of object* and *exploitation event*.

During the object exploitation process, i.e. during the process of object utilization and maintenance, the properties of an object change. For some features, these changes will be of a continuous type, sometimes slow, sometimes transitional, and sometimes drastic. Therefore, an object at a given moment in time is not identical to the object at a different moment in terms of its properties. In order to describe the process of these changes the term *state* is applied.

Defining a set of object properties $\mathfrak{C}$, $\mathfrak{C} = \{c_1, c_2, \ldots, c_m\}$ the state of the object in a time t is determined by the function:

$$\mathfrak{S}(t) = f\,[\mathfrak{C}(t)] = f\,[c_1(t), c_2(t), \ldots, c_m(t)].$$

Kaźmierczak (2000 p. 119) gave a similar assessment of the term 'state': 'under the term state of object we are going to understand here a "photography" of values of object properties in a given moment of time'.

In practical applications, this function is not considered a continuous one. Discretization is usually made and states are named. These names are usually associated with the physical side of the state, e.g. repair state, work state, standstill state, etc. Notice that a simple conclusion can be made here: the exploitation process of an object is the sequence of states of this object or—the usual formulation—this is the process of changes of states.

As a result of this discretization to each moment when a change of state appears, an exploitation event has taken place. Sometimes these events are *visible* and to some extent *perceptible*, e.g. a certain element of the object failed and the machine ceases operation. Sometimes events are conventional ones—nothing physically happened apart from the fact that a certain object parameter exceeded its limited value, e.g. brake lining worn excessively. In this moment, it is assumed that the object is in a different state.

The **process of changes of reliability states** of shovels is identified here as the alternative process—work-repair (Figure 5.1)—whereas for transporting machines an additional state—reserve—has to be included, and the sequence here is work—repair—reserve (Figures 2.1 and 2.2).

The **process of functioning** is more complicated.

For shovels realizing their loading task, it is important whether the shovel is capable of loading or not. It is obvious that during the repair state loading is impossible. But there are some periods during a work state when a machine can load but does not do so because it is being used to carry out a different operation, e.g. it moves to a new loading point. Thus the shovel is *inaccessible* for loading. By considering the problem of accomplishing a loading task by a shovel, these two states can be lumped (agglutinated, joined together)[5] to obtain one—a state of *incapability* for loading.

Here are all the distinct states of a shovel:

$\mathfrak{R}_{np}$—repair state (failure clearing)
$\mathfrak{R}_{nd}$—state of inaccessibility for loading
$\mathfrak{R}_{nz}$—state of incapability for loading
$\mathfrak{R}_{p}$—work state
$\mathfrak{R}_{zd}$—state of ability (and also accessibility) for loading.

Such a list is called an *exploitation repertoire*.

The relationships between states are as follows: (Figure 5.2)

$$\mathfrak{R}_{np} \cup \mathfrak{R}_{nd} = \mathfrak{R}_{nz} \qquad \mathfrak{R}_{zd} \cup \mathfrak{R}_{nd} = \mathfrak{R}_{p}$$

$$\sim \mathfrak{R}_{np} = \mathfrak{R}_{p} \qquad \sim \mathfrak{R}_{p} = \mathfrak{R}_{np}$$

$$\mathfrak{R}_{zd} = \sim(\mathfrak{R}_{np} \cup \mathfrak{R}_{nd}) \qquad \mathfrak{R}_{np} \cup \mathfrak{R}_{p} = \mathfrak{R}_{nz} \cup \mathfrak{R}_{zd}.$$

It appears at first glance that all possible shovel states are enumerated. But this is not the case. One unwanted state would probably occur when a shovel starts its cooperation with the haulage means, i.e. when it becomes an element of a certain system. This state is a standstill state $\mathfrak{R}_{s}$, i.e. the machine is waiting idly for a truck. The shovel is in a work state (reliability state) and also in a state of accessibility for loading (exploitation state) but does not load because of the lack of a transporting machine. The frequency of occurrence of this state depends on the organization of the whole machinery system, the number of elements of a particular type in the system and the geological and mining conditions of the mine. This statement will be proved further in this book. The existence of this state will be taken into consideration during modelling (Figure 7.6). The expression *shovel work cycle* is also connected. This is a measure of shovel actions made in time that are associated with loading; actions repeated periodically.

---

[5] In mathematics in the theory of states, mathematicians often lump some states (see, for example, Zakharin 1972, Markovsky and Trcka 2006, Callut and Dupont 2004).

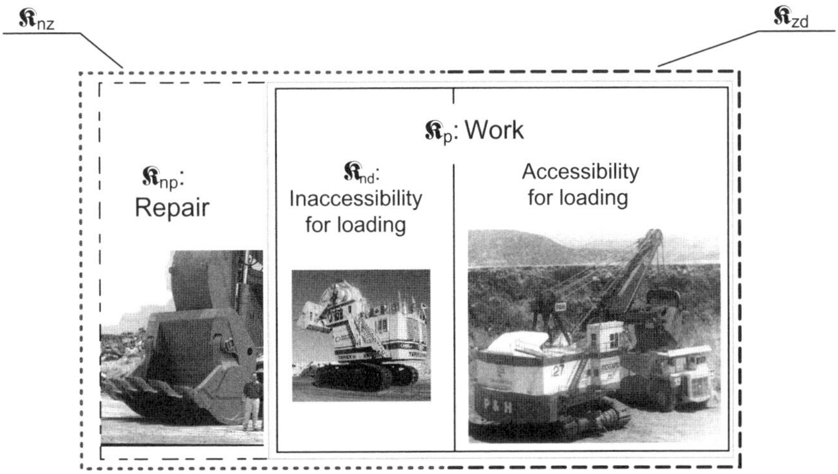

Figure 5.2. Relationships between shovel exploitation states.

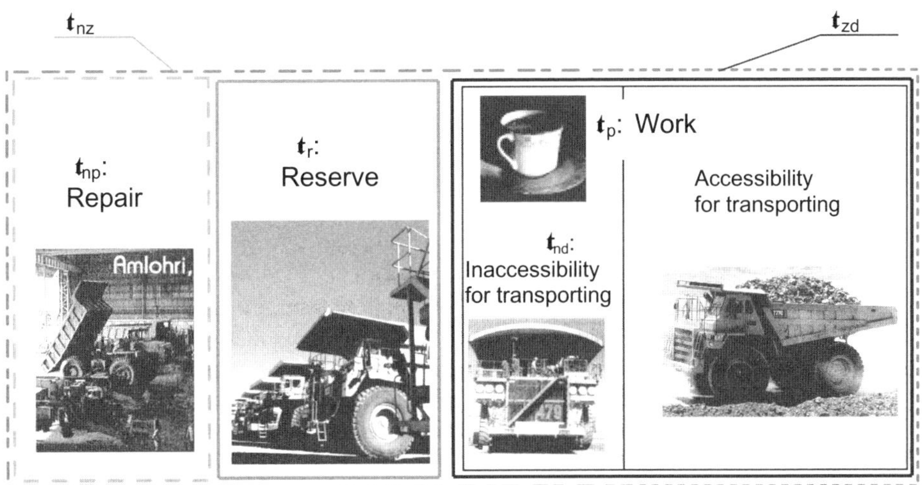

Figure 5.3. Relationships between truck exploitation states.

Considering the process of the functioning of trucks, their accessibility for conducting transportation and their inaccessibility will also be looked at. The exploitation repertoire for trucks is similar to that for shovels. The only difference is one state more—reserve. Thus, a truck exploitation repertoire is as follows:

$\mathbf{t}_{np}$—repair state (failure clearing)
$\mathbf{t}_{nd}$—state of inaccessibility for transporting
$\mathbf{t}_{nz}$—state of incapability for transporting
$\mathbf{t}_{p}$—work state
$\mathbf{t}_{r}$—reserve state
$\mathbf{t}_{zd}$—state of ability (and also accessibility) for loading. (Figure 5.3)

As with shovels, a standstill state $\mathbf{t}_{s}$—the truck waits in a queue—also needs to be included. Again, this state is a by-product of the fact that the truck operates in a system. By looking more carefully at truck system operations in mines, it can be seen that a queue can be formed at any point on the truck route and also that a truck can be in a queue waiting for repair. In this consideration, attention is paid to the most frequent cases when, first, the truck is in a queue waiting for repair and, second, it is waiting before the shovel for loading. The possibility of these queues appearing will be considered during modelling, employing appropriate queue models (Figures 2.2 and 7.7).

The expression *truck work cycle* is also connected with trucks. It is a time measure of the sequence of truck actions that are associated with transportation; actions repeated periodically. Here there are four stages of this cycle: load—haul—dump—return (Figure 9.1).

CHAPTER 6

Probabilistic properties of components of the machinery system exploitation process

This chapter gives an idea of the kinds of distributions that are used to describe the dispersion of values of random variables, components of the modelling procedure. A short review of the literature available in this field is given together with the results obtained during the author's research.

6.1 SHOVEL REPAIR TIMES

Although the distributions of shovel repair times were recognized many years ago, the number of publications to date can hardly be described as rich. Typical histograms of repair times based on gathered statistical data are shown in Figures 6.1 and 6.2. The first diagram was created by Temeng (1988) during his M. Sc. dissertation. Czaplicki (1986–1988) produced the second one.

The literature on the subject—Czaplicki (1986–1988), Temeng (1988), Czaplicki and Temeng (1989), Kolonya *et al.* (2003)—is unanimous—theoretical distributions that describe well the empirical data of shovel repair times are gamma or Weibull distributions. A similar conclusion can be formulated by analyzing the data given by Nanda (2003).

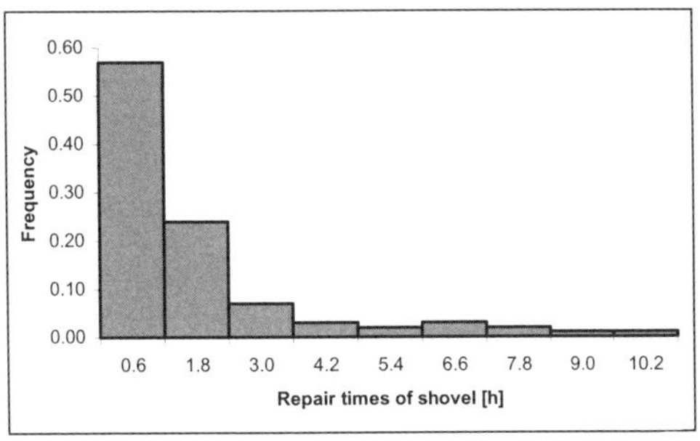

Figure 6.1. Shovel repair times histogram—Temeng (1988).

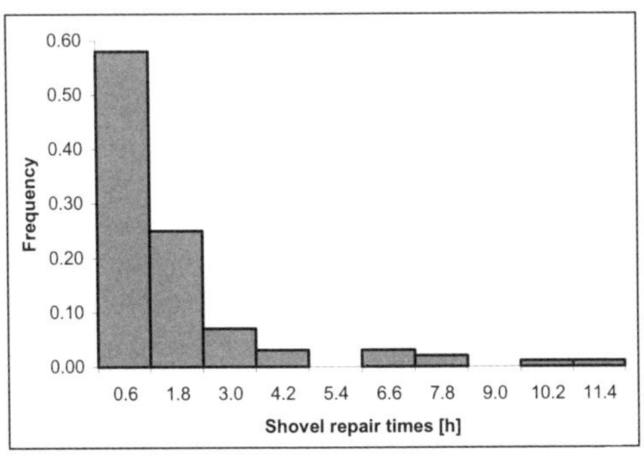

Figure 6.2. Shovel repair times histogram—Czaplicki (1986–88).

6.2 SHOVEL WORK TIMES

Histograms of work times between two neighbouring repairs are abbreviated to histograms of work times. For shovels, they are presented in Figures 6.3–6.5.

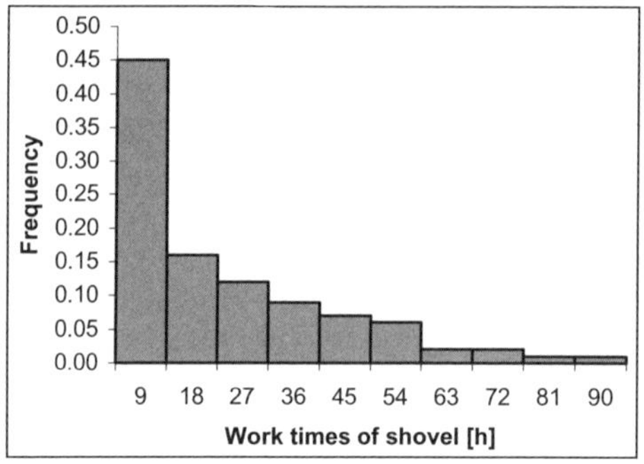

Figure 6.3. Shovel work times histogram—Temeng (1988).

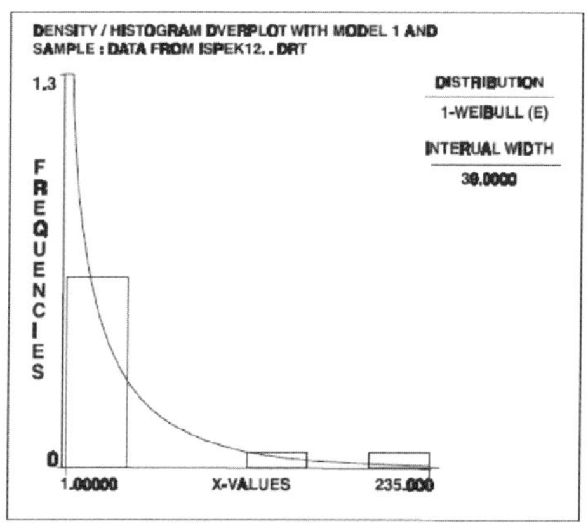

Figure 6.4. Shovel work times density function—Kolonja *et al.* (2003).

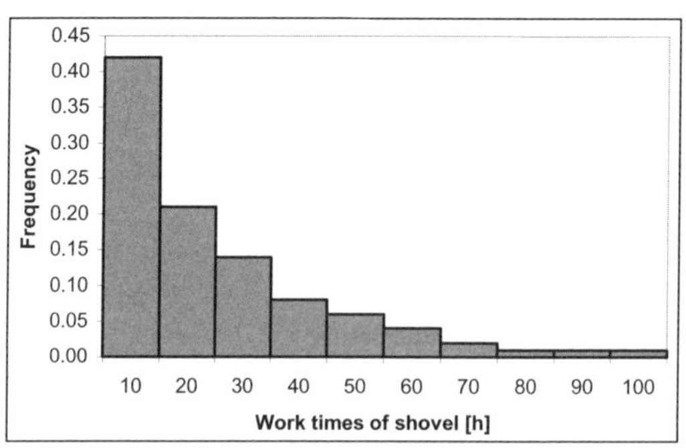

Figure 6.5. Shovel work times histogram—Czaplicki (1986–88).

They are distinctly asymmetric. According to the investigations made by Temeng 1988, Kolonya *et al.* (2003) and Czaplicki (1986–88), the theoretical distributions that describe empirical data well are again Weibull or gamma. However, the distributions of repair times are more often described by these distributions with shape parameters different than 1, whereas work times can usually be satisfactory described by these distributions with a shape parameter almost equalling 1. Therefore, it can be assumed as the exponential.

An applied test of goodness of fit in the majority of cases gave no grounds to reject the statistical hypothesis stating an exponential distribution. In all statistical investigations made by Temeng and Czaplicki the level of significance was assumed to be 0.05.

6.3 TRUCK REPAIR TIMES

The majority of histograms of truck repair times are asymmetric in character (Figures 6.6 and 6.7). After more careful statistical analysis, though, differences in this character can be found. For this reason, theoretical distributions, either gamma or Weibull, have shape parameters below, equal to or more than 1. Temeng (1988) proved that the character of these distributions depends—among other things—on the participation of different types of repairs in the whole data.

Temeng divided the whole repair data into elementary failure clearing and compound clearing and, additionally, into electrical and mechanical types of repairs. Histograms of elementary repair times and electrical repair times were almost always asymmetric, accurately described by an exponential probability distribution. Histograms of mechanical repair times and compound repair times were asymmetric, frequently positive asymmetric (Temeng 1988 pp. 59–62). Czaplicki and Temeng (1989) proved that different types of trucks have different probability distributions describing their repair times, but gamma or Weibull distribution can still be employed. The difference is visible in the values of the function parameters. From a physical point of view, these differences are associated with the different constructional solutions of particular assemblies. The difference can be traced even when comparing the same trucks made by one producer where one machine has a new version of a particular assembly and the second has the old solution.

To conclude these considerations it should be added that Nanda (2003) presented an assessment of the reliability parameters of selected assemblies of shovels and trucks cooperating with them, while Mrig (1991) demonstrated an analysis of the reliability of machinery equipment operating in surface coalmines of South-East India. The failure problems of drillers, draglines, power shovels and trucks were all described.

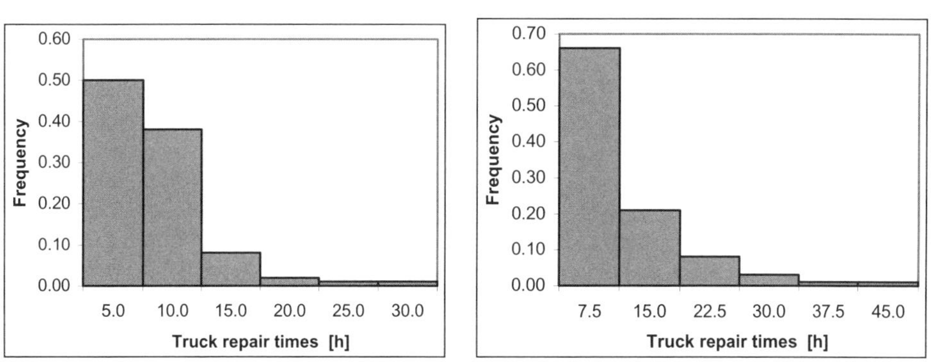

Figures 6.6 and 6.7. Truck repair times histograms—Czaplicki (1986–88).

6.4 TRUCK WORK TIMES

The probability distributions of the work times of trucks can, in the majority of cases, be satisfactory described by exponential distributions (Figures 6.8–6.10). Temeng (1988) proved during his

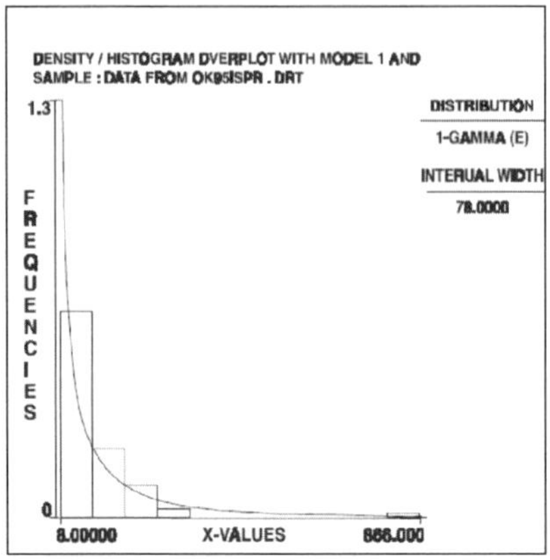

Figure 6.8. Truck work times density function—Kolonya *et al.* (2003).

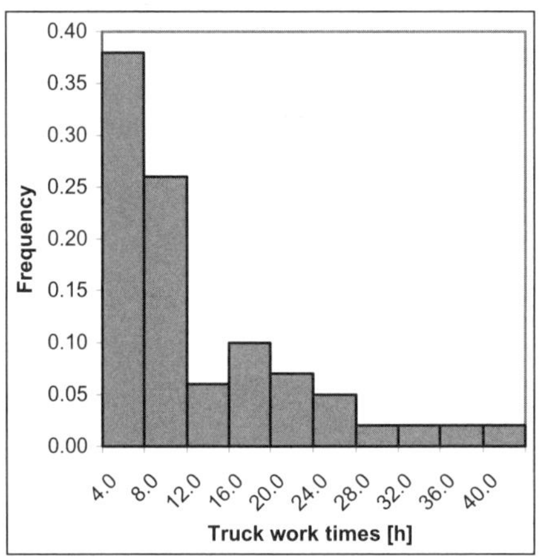

Figure 6.9. Truck work times histogram Temeng (1988).

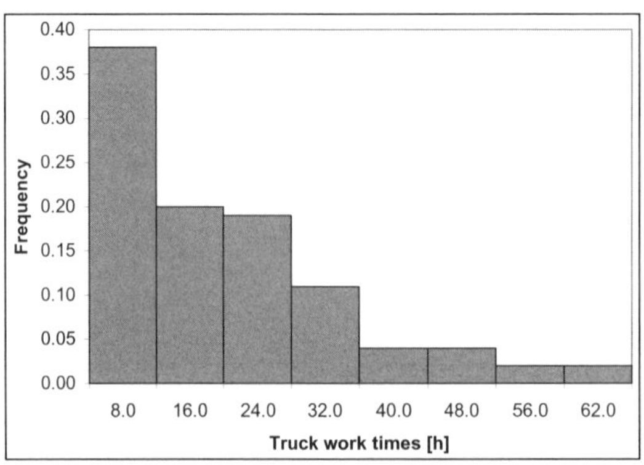

Figure 6.10. Truck work times histogram—Czaplicki (1986–88).

research that all trucks investigated—44 vehicles were Haulpack 120-ton, 14 LectraHaul 120-ton and 26 LectraHaul 100-ton—possessed truck work times distributions that could be described by exponential probability distributions.

Czaplicki obtained similar results during extensive investigations carried out from 1986 to 1988. The system observed operated in the Nchanga Open pit in the Copperbelt, Zambia.

Kolonya *et al.* (2003) also employed gamma distribution and Weibull distribution to describe the empirical data of work times of trucks, even though the shape parameters were quite close to unity.

In Carmichael's monograph (1987) repair times of trucks and work times of trucks were depicted by exponential distributions.

6.5 TIMES OF TRUCK WORK CYCLE PHASES

The set of functions applied to describe the loading times of a truck by shovel is quite rich, starting from exponential probability distribution—that is not real, but employed because of its convenient properties (e.g. Barbaro and Rosenshine 1986, Barnes *et al.* 1979, Morgan and Peterson 1968, Nanda 2003, Panagiotou 1993, Purohit and Nanda 1995); through Weibull function (Lizotte *et al.* 1989); also with a third, displacement parameter (Lizotte and Bonates 1986, Temeng 1988); logarithmic-normal function (Griffin 1989, Hufford *et al.* 1981); Erlang function (Zhongzhou and Qining 1988); up to Gaussian function (Kolonya *et al.* 2003, Wright 1988). Samples of histograms constructed based on empirical data gathered by Czaplicki (1986–88) and Temeng (1988) are presented in Figures 6.11–6.14.

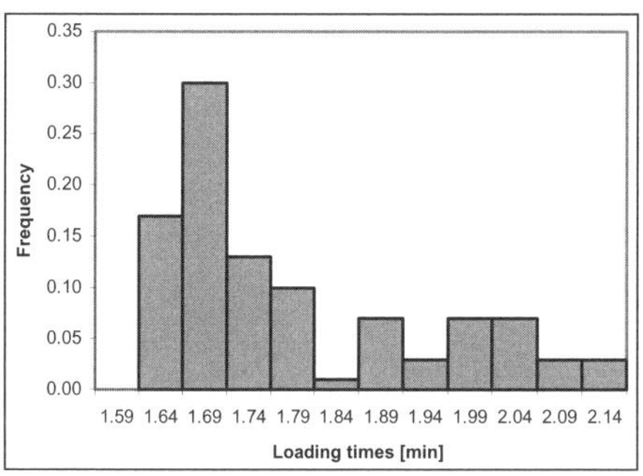

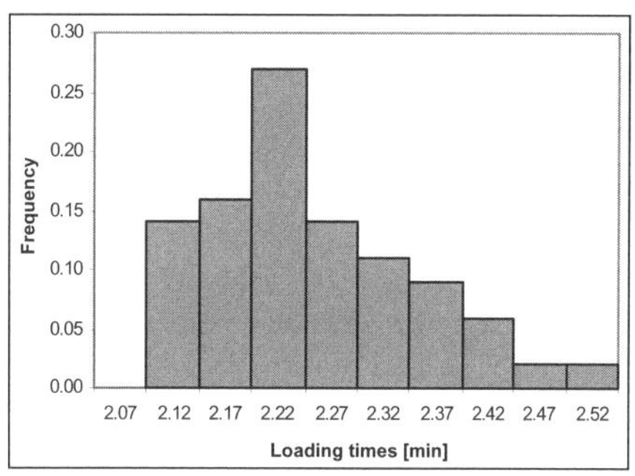

Figures 6.11 and 6.12. Truck loading times histogram—Temeng (1988).

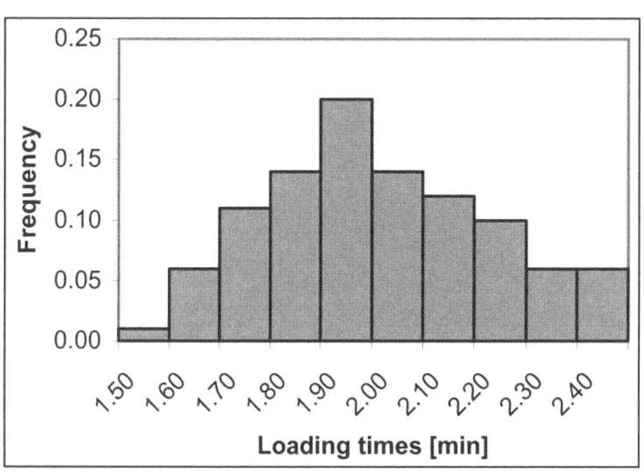

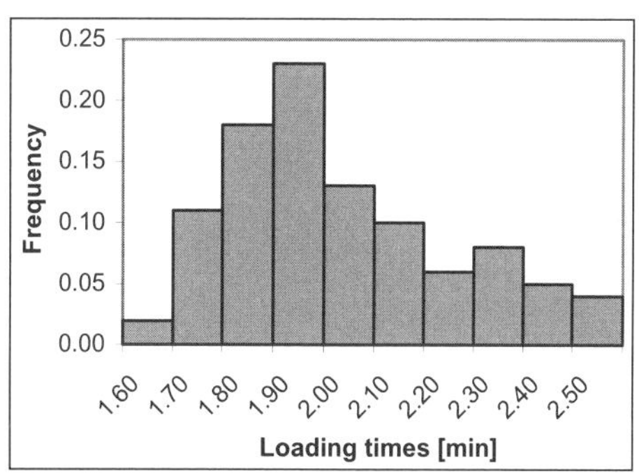

Figures 6.13 and 6.14. Truck loading time histogram—Czaplicki (1986–88).

One subtle problem is connected with the phase of loading. A truck that has just arrived at a shovel which is free and able to load usually spends some time manoeuvring in order to place its empty box correctly and conveniently for loading by the shovel. This takes some time. This time is called *spotting time*, or *time of truck replacement* (Koziol and Uberman 1994). Some authors include this time, adding it to the time of pure loading, while others separate it, dividing loading into two phases (Panagiotou 1993, Figures 6.15–6.17). It makes no sense to label either approach the 'proper' one. Both ways are correct, depending on the purpose of the investigation. For the purposes of discussion in this monograph, there is no need to create an additional phase. Loading is an action between two machines, thus spotting time should not be included in the loading phase.

Based on my own research and making use of a review of the appropriate literature, an application of exponential probability distribution to describe dispersion of loading times is not adequate for at least two reasons. First, the distribution of the probability mass over the values of random variable is diametrically different from the distribution of the empirical mass shown in the histograms. Second, the standard deviation of the variable in practice ranges between 0.2 and 0.4, rarely exceeding 0.5 in relation to the mean value. But the regularity of exponential distribution is that the variation coefficient is 100%, that is the mean value equals the standard deviation. These two disagreements negate the point of applying exponential probability distribution. Employment of logarithmic-normal distribution sometimes gives a fairly good approximation, and sometimes a poor one.

If descriptions of loading times by Weibull, Gauss and Erlang distributions for the same set of data are compared, say: mean loading time $\overline{x} = 2.1$ min and standard deviation $s = 0.6$ min. These density functions are shown in Figure 6.18.

It is easy to trace the pattern that the Gauss and Weibull distributions have, as they are an almost identical plot. The Erlang function is slightly different, first because the shape parameter, 12.25, has been rounded up to the nearest natural number—12, according to the Erlang distribution requirements. Thus, the application of this probability distribution makes sense when the properties of these functions will be useful in further considerations, e.g. when the Erlang type system is being analyzed. However, these systems have their own disadvantages (Czaplicki 2004).

The unloading times are shorter compared to dumping times because this is unit technical action. This time does not depend on the number of buckets loaded. All these mean that the dispersion of these times is well described by normal distribution (Figure 6.19).

Similarly, the times of material haulage and truck return to loading shovel may be described successfully by Gauss probability distribution (Figures 6.20–6.22).

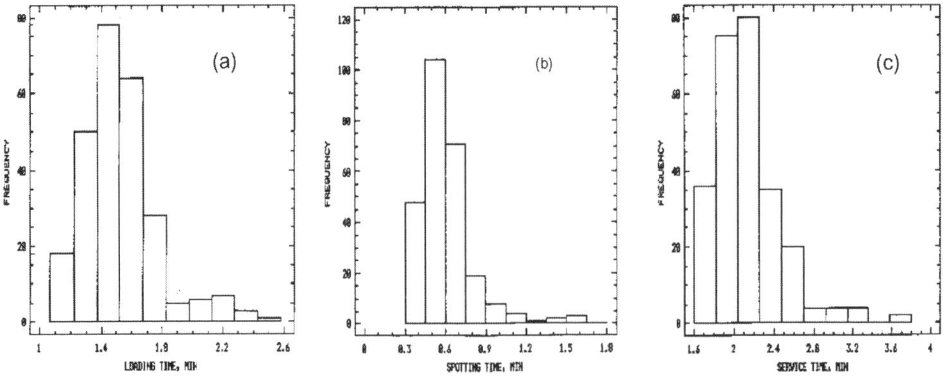

Figures 6.15–6.17. Histograms: pure loading time (a), spotting time (b), total time (c)—Panagiotou (1993).

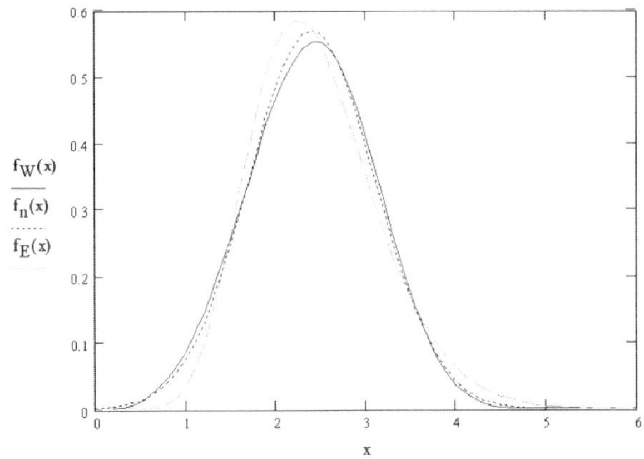

Figure 6.18. Probability density functions of truck loading time: $f_W(x)$—Weibull's distribution, $f_n(x)$—normal distribution, $f_E(x)$—Erlang's distribution, for the same mean and standard deviation.

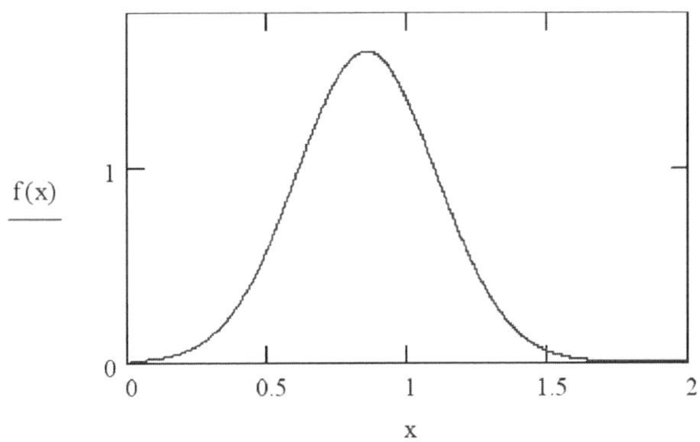

Figure 6.19. Truck unloading time density function—Wright (1988).

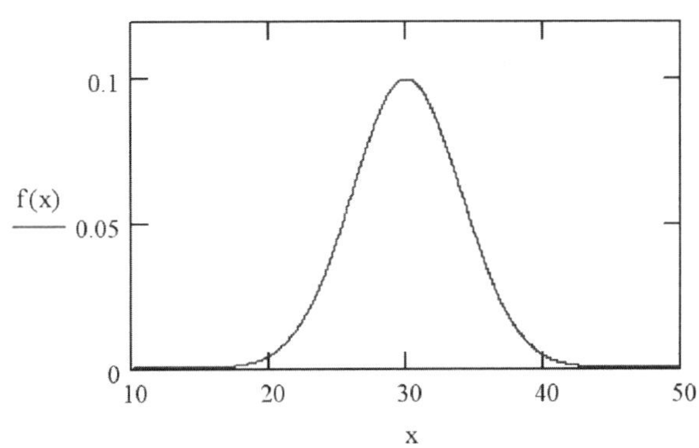

Figure 6.20. Truck return time density function—Wright (1988).

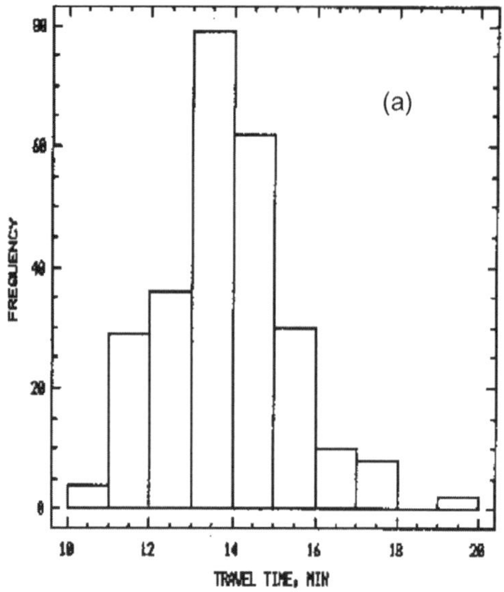

Figure 6.21. Truck haulage time density function—Panagiotou (1993).

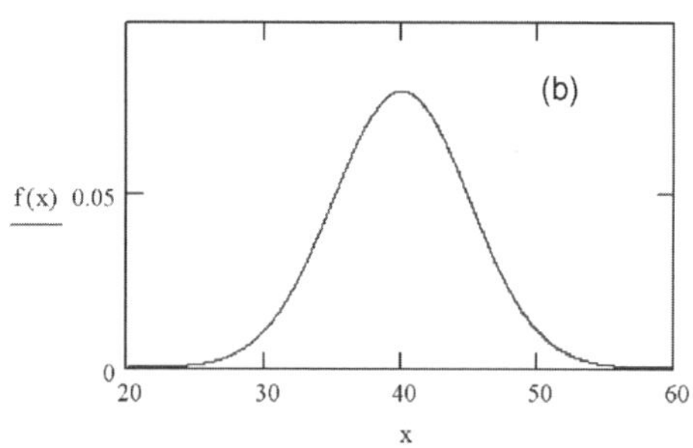

Figure 6.22. Truck haulage time histogram—Wright (1988).

To recap. It is now clear what kind of random variables can be employed to describe the dispersion of times of both random variables which are components of the reliability of machines involved in the system. The distribution of times of components in the truck work cycle is also known. This information is a basis for the selection of appropriate queuing models and will be required in further considerations.

CHAPTER 7

Modelling and analysis of the exploitation process of a shovel-truck system: Part I

Analytical models are widely applied in the world of mining engineering to analyze and assess machinery systems of the shovel-truck type based on models of queuing theory. A significant characteristic of these models, according to, among others, Kopocińska (1968), Barnes *et al.* (1979), Barbaro and Rosenshine (1986), Carmichael (1987) and Fabian (1989), is that the number of operating machines is assumed to be constant. This is contrary to operational practice. Czaplicki (2002, 2004, 2004/2005 and 2006) presented a different solution to the problem of the machinery system size accomplishing its task in a given moment of time. In the cited papers, the following is emphasized repeatedly: 'the number of machines in work state is a random variable'. There is no doubt that the identification of factors influencing the probability distribution of this random variable is interesting from both theoretical and practical points of view. As Czaplicki states (2006) 'identification of the probability distribution of the number of machines in work state gives a basis for trustworthy research—analysis and estimation of efficiency[1] of the machinery system. This probability distribution has an influence on the values of most measures of system performance.' Therefore, when commencing modelling in this chapter, the focus will be on the construction of this probability distribution. During modelling, various properties of the system will be investigated, creating a comprehensive analysis of it.

7.1 SYSTEM OF SHOVELS

In open pit mines, power shovels are used to load excavated rock on to trucks. In the majority of cases, this material is attained by blasting, but some mines also employ a shovel-truck system to excavate bedded type deposits such as coal seams. This excavation is generally carried out using bulldozer-rippers, supported by blasting, where necessary.

In most cases, power shovels have the same or almost the same bucket/dipper capacity. This solution is very convenient because trucks can be directed to any loading machine. Similarly, the best solution is when transporting machines have the same payload. This type of solution is assumed to be the basis of further considerations within this monograph.

The first step for modelling of the shovel-truck system is the analysis of the shovel subsystem. It is assumed that this system has been properly selected according to exploitation needs. The system generates—borrowing a term from mathematics—the stream of mineral that is a transportation task for the hauling subsystem.

The number of operating shovels is denoted by n. Looking at this system from the point of view of reliability, it can be seen that it is a system of n machines working independently of each other.

---

[1] It can sometimes be a problem to distinguish between the terms: efficiency, effectiveness, effectivity and efficacy. We presume here that the term efficiency will be understood as: 'the quality or property of being efficient'. We define this word as 'a system feature that characterizes the degree to which the system abilities have been used in the process of achieving a given goal in determined conditions of this realization' (Sienkiewicz 1987).

Still keeping reliability in mind and applying elementary reliability principles the probability distribution of a number of machines is determined as—for instance, in repair:

$$P_{kd}^{(n)} = \binom{n}{d}\left(\frac{1-A_k}{A_k}\right)^d P_{k0}^{(n)} \qquad d = 1, 2, \ldots, n \tag{7.1}$$

whereas

$$\sum_{d=0}^{n} P_{kd}^{(n)} = 1$$

where: $P_{kd}^{(n)}$—probability that d shovels are in repair state[2]
A_k—steady-state availability of the shovel.

Denoting by ε the intensity of shovel failures and by η the intensity of shovel repairs, the steady-state availability of a shovel is defined by the formula:

$$A_k = \frac{\varepsilon}{\varepsilon + \eta}. \tag{7.2}$$

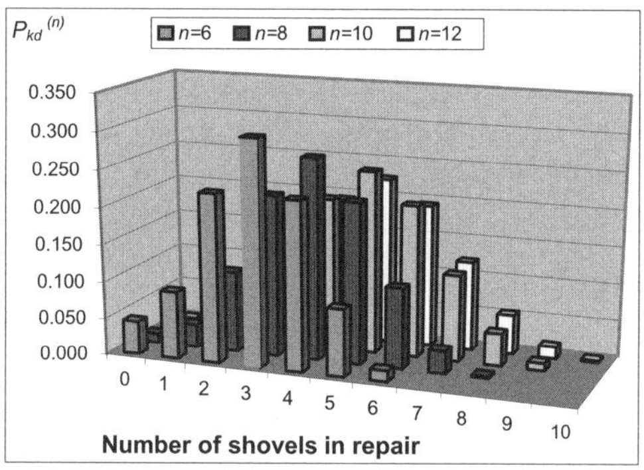

Figure 7.1. Probability distributions $P_{kd}^{(n)}$ of numbers of shovels in repair for different numbers n of shovels in the system for the steady-state availability $A_k = 0.750$ (medium reliability).

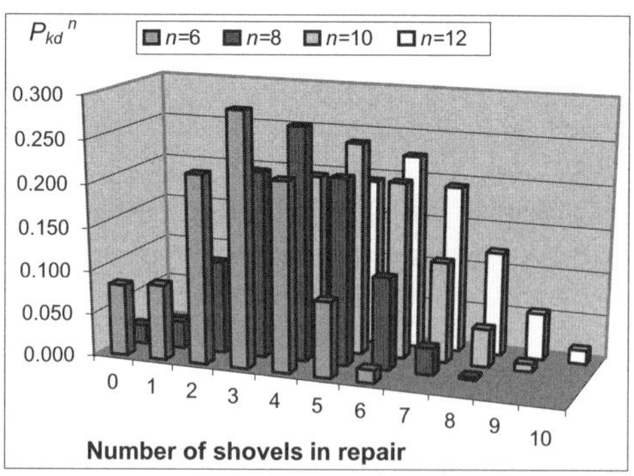

Figure 7.2. Probability distributions $P_{kd}^{(n)}$ of numbers of shovels in repair for different number n of shovels in the system for the steady-state availability $A_k = 0.850$ (high reliability).

[2]All parameters connected with shovels will be marked by k, connected with trucks by w.

The probability distributions for a number of shovels in the repair state for different number of shovels $n = 6, 8, 10, 12$ in the system for two different levels of their availability $A_k = 0.750$ and $A_k = 0.850$ are shown in Figures 7.1 and 7.2.

Similarly, the probability distributions for a number of shovels in the repair state for different levels of availability $A_k = 0.650, 0.750, 0.850$ for $n = 6$ and 12 shovels are shown in Figures 7.3 and 7.4.

The expected number of shovels in repair is given by the formula:

$$E_k^{(n)} = n(1 - A_k) \tag{7.3}$$

The exploitation process of the shovel relies on loading the arriving trucks, carried out many times until the moment when the blasted material is entirely cleared away from a given place. The shovel then moves to a new loading place where recently blasted rock waits for removal. It has been assumed (Church 1981, Czaplicki 1997, 2004, Dudczak 2000) that all additional operations made by a shovel—except for loading—are divided into long-lasting operations (e.g. the machine moves to a new loading point, moves out from the face allowing a blast, etc.) and short-lasting operations. It can be assumed that these short operations do somewhat extend the loading time. These small operations do not stop the process of the arrival of trucks at the loading shovel. According to estimates made by Church (1981) and Czaplicki (1989) the time lost due to these reasons can be assumed to be 5%. Thus, for further reasoning it is assumed that the mean loading time is given by the formula:

$$Z' = 1.05Z$$

where Z is the mean loading time of the truck. The magnitude Z' is named as *the mean adjusted loading time of the truck*.

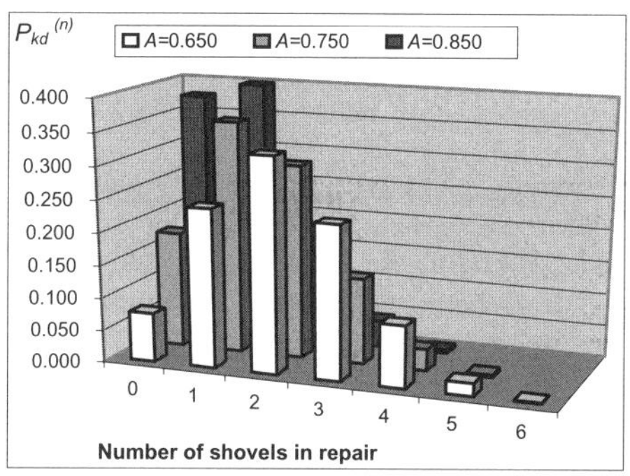

Figure 7.3. Probability distributions $P_{kd}^{(n)}$ of numbers of shovels in repair for different steady-state availability $A_k = 0.650, 0.750, 0.850$ for 6 shovels.

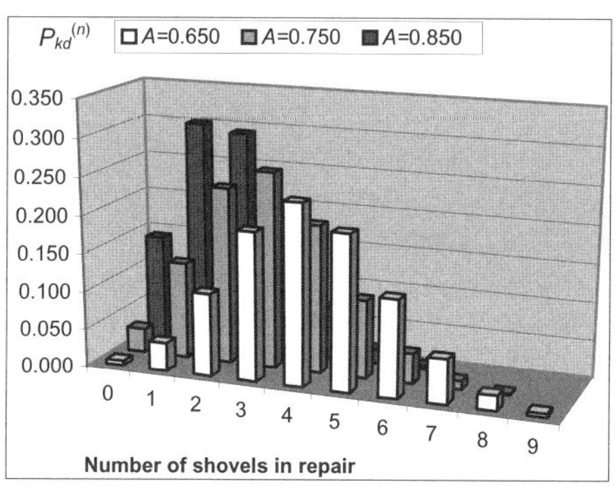

Figure 7.4. Probability distributions $P_{kd}^{(n)}$ of numbers of shovels in repair for different steady-state availability $A_k = 0.650, 0.750, 0.850$ for 12 shovels.

Long-lasting operation of a shovel causes the truck dispatcher to direct empty haulers to go down the pit to the remaining loading machines, excluding for the time being this shovel from the dispatching scheme. This means that the accessibility of the shovel shortens. In the past, the approach to the problem of estimating accessibility was different (e.g. Caterpillar Performance Handbook 1996, Church 1981, Surface Mining SME 1990, SME Mining Engineering Handbook 1992, Terex Manual 1981). Some publications give a gradation of the accessibility by a special coefficient depending on different factors. This gradation is frequently applied in mine engineering practices for the calculation of system productivity or the estimation of system efficiency. For example, for *good* organization of shovel work the accessibility coefficient B_k is assessed to be 0.85, meaning that 15% of time is used for operations other than loading.

However, it should be realized that many different factors influence the estimate of this coefficient, and sometimes it is hard to imagine that the diverse operations of the mine decide its value. The following are some of the directions required for the improvement of the coefficient value:

- Avoid excessive shovel moves
- Maximize production at the lowest number of benches and minimum number of working areas
- Expose material for excavation in a timely fashion
- Assure adequate working room
- Assure proper change of shovel operators.

The final characteristics of a shovel system can now be determined.

Taking into account that:

a. Shovel states of repair and accessibility for loading are independent of each other
b. Shovels operate independently of each other,

the probability distribution of the number of shovels in the state of accessibility for loading can be constructed. This distribution is given by:

$$P_{kd}^{(zd)} = \binom{n}{d} G_k^d (1 - G_k)^{n-d} \quad d = 1, 2, \ldots, n \quad G_k = A_k\, B_k \tag{7.4}$$

where $P_{kd}^{(zd)}$ is the probability that d shovels are in the state of accessibility for loading, i.e. d shovels are able to load and G_k is the shovel loading capability coefficient.

Figure 7.5 shows an example probability distribution $P_{kd}^{(zd)}$ for a system of 8 shovels of the steady-state availability $A_k = 0.800$ and the accessibility coefficient $B_k = 0.850$.

The exploitation graph[3] for shovels can be illustrated as in Figure 7.6.

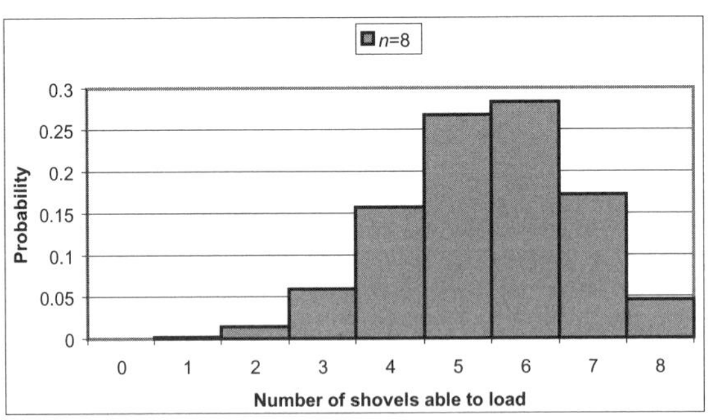

Figure 7.5. Probability distribution of numbers of shovels in state of accessibility for loading for a system of $n = 8$ shovels of the steady-state availability $A_k = 0.800$ and the accessibility coefficient $B_k = 0.850$.

[3]Term 'graph' used here does not mean 'picture' but is taken from mathematical theory of graphs. (see for example Gould 1988, Merris 2000). Graphs are widely applied in reliability and exploitation theories.

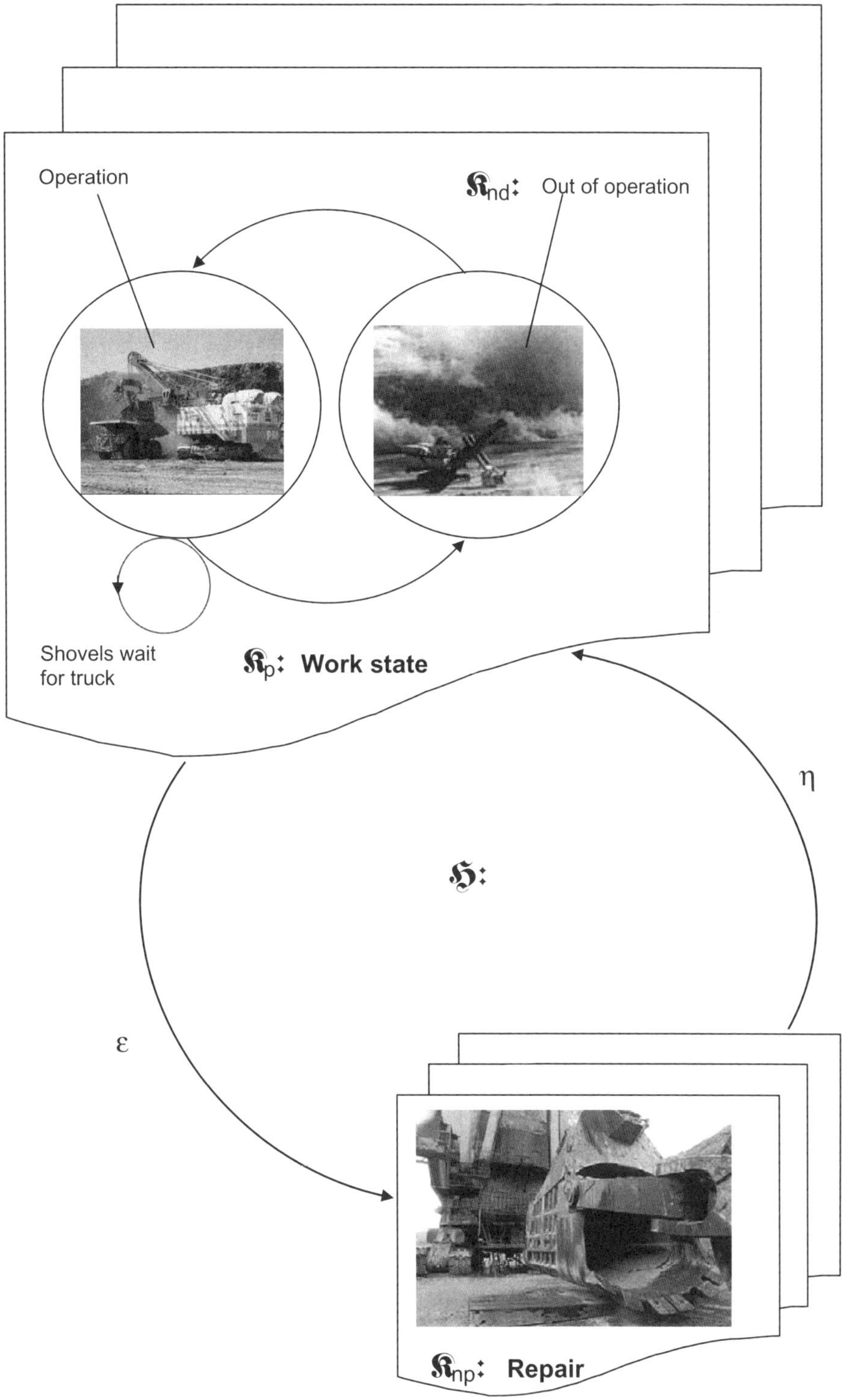

Figure 7.6. Exploitation graph for shovels.

7.2 TRUCK-WORKSHOP SYSTEM

The second system to be considered, but also the largest subsystem of the shovel-truck system, is the fleet of hauling machines. Reliability of these transporting units and an appropriately organized back-up facility for repairs, overhauls, technical surveys, etc. have a great influence on the efficiency of the whole machinery system and the number of trucks accomplishing their transportation task at any given moment.

The point of consideration in this chapter will be the construction of the probability distribution of a number of trucks in the work state as the function of the reliability of trucks, number of repair stands and intensity of truck repairs.

If a shovel-truck system is small—say, one or two loading machines and several or a dozen trucks—then the Maryanovitch model can be applied to analyze it. The reason for this is the fact that in small systems the size of the repair shop could be relatively large, with the possibility of repairing all units in failure simultaneously. There will not be a queue of trucks waiting for repair and the Maryanovitch model gives a precise estimation of the number of machines in repair.

At present, in the majority of machinery systems of this type up to a dozen or so shovels cooperate with a few dozen or sometimes over a hundred trucks. For such a system, the back-up facility to maintain a fleet of machines does not have such a great number of repair stands to mend all failed units at any moment. Such large repair equipment does not make economical sense. In addition, these days a number of mines operate in mountains. The construction of a huge maintenance bay is, in such a case, extremely expensive, requires time, and usually engages a lot of machinery. On the other hand, the probability of an event where all transporting units are out is tremendously low for large systems.

For a system consisting of m trucks directed to haul excavated material and r trucks in a cold reserve. Machines in a work state can fail with the intensity δ and the repair is done with the intensity γ. The number of repair stands is k. The standard deviations of work times and repair times are known and are marked by σ_p and σ_n, respectively. If so, the following parameters can be calculated:

$$C_M = (\delta\sigma_p)^2 \quad C_R = (\gamma\sigma_n)^2 \quad \xi = \delta/\gamma. \tag{7.5}$$

The first two parameters[4] are the square coefficients of the variation of uptimes of operating machines and repair times respectively. The third parameter is the failure rate of trucks in work state.

Figure 7.7 shows a scheme of machine flow in the system with the intensities of transition between states and places of possible queues.

In the system is a reserve of trucks.

The majority of manuals, textbooks, handbooks (e.g. SME Mining Engineering Handbook[5] 1973, Terex Manual 1981, Hartman 1981, (Polish) Mining Engineering Handbook 1982, Caterpillar Performance Handbook 1996, Vergne 2003 and Czaplicki 2004) state that the truck reserve size depends—first of all—on the number of units needed to accomplish the formulated transportation task. However, recommendations how to assess that size are determined in various ways.

Hartman (1981) gave the following recommendation (p. 256): 'Spare units. To maintain a full haulage fleet in operation even when breakdowns occur, spare units are usually purchased. For every five to six production units at the mine, one spare is provided'.

The Terex Manual (1981) offers the advice (p. 48) that the system of trucks should be filled up by the number of transporting units lost due to the unreliability of the truck fleet.

---

[4] In the further considerations the first coefficient of variation will be assumed as $C_M \approx 1$ because the probability distribution of times of work state of trucks is in most cases exponential.

[5] It is strange that the problem of truck reserve is non-existent in Surface Mining SME (1990) as well as the Mining Engineering Handbook (1992).

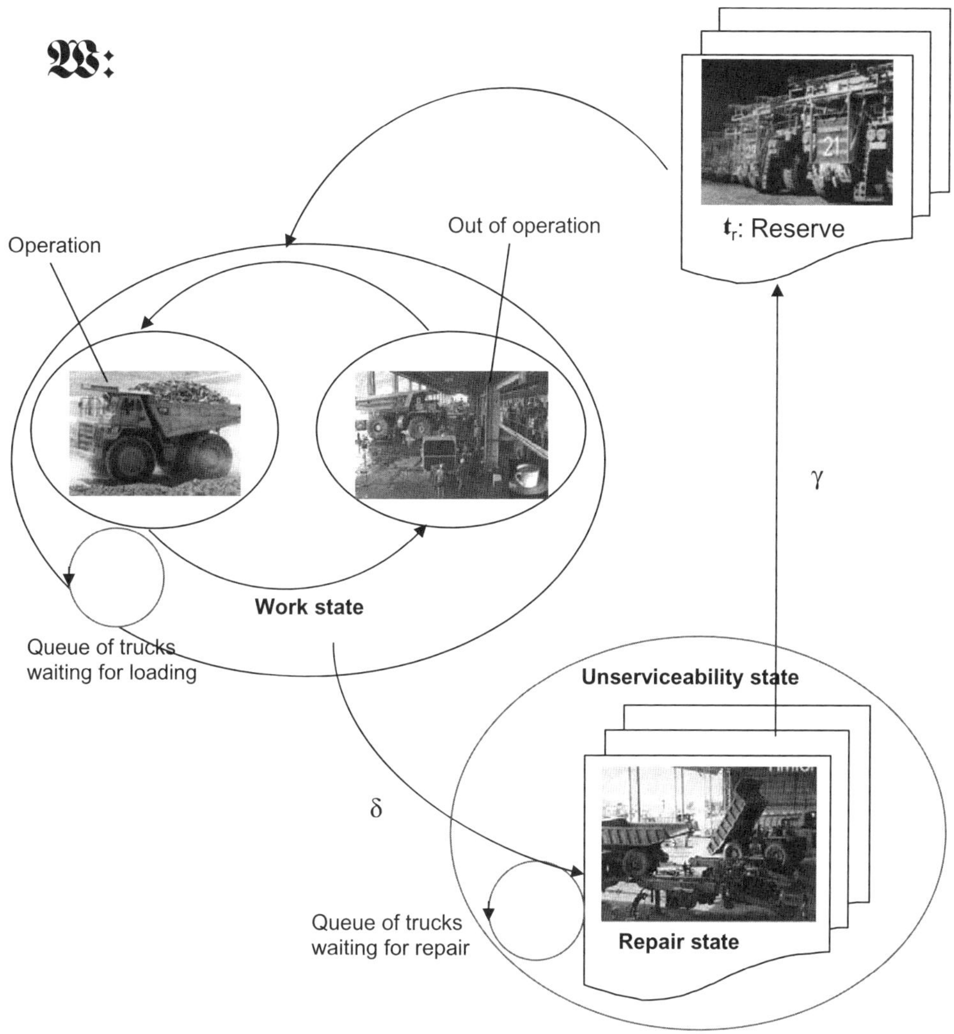

Figure 7.7. Exploitation graph of trucks.

Both Mining Engineering Handbooks—SME (1973) and the Polish (1982)—recommend (pp. 18.21–18.23 and p. 296 respectively) thc same—that the size of the truck fleet should be defined by the quotient of the number of trucks needed to accomplish the transportation task to the steady-state availability of the truck.

It is easy to prove (see for instance Czaplicki 1992) that following these recommendations we can obtain an advisable numbers of spares with a dispersion of over 100%.

This result was a motivation to approach the problem of spares in a different way. In the papers cited above (Czaplicki 1992 and 2006) the reserve size was determined as the result of the division of the number of trucks required to accomplish the formulated transportation task into a pair of numbers $< m, r >$ in such a way that the number of trucks in the work state must not be less than that obtained from the appropriate calculation. The criterion of the pair selection has two postulates: the total number of trucks in the system should be the lowest possible and the reserve size the largest possible.

Czaplicki (2006) proved that for the number of parameters considered m, r has to be extended by the number of repair stands k. This proof will be given later. The parameters m, r, k will hereafter be called *the structural parameters of the shovel-truck system.* However, these parameters concern a truck-workshop system.

As far as the recommended number of repair stands for trucks is concerned, relevant literature on this topic is almost non-existent. Some suggestions of truck producers can be found, but these hints are published without any scientific evidence.

The following are some preliminary considerations in connection with the number of repair stands required for a given truck fleet.

As noted by Czaplicki (2006), is it advantageous when the following inequality holds:

$$k > r \tag{7.6}$$

Taking into the account inequality (2.11) the following can be written:

$$k > m\frac{\delta}{\gamma} \tag{7.7}$$

whereas the heavy traffic condition calls for

$$\frac{m\delta}{k\gamma} \geq 0.75.$$

When the steady-state availability of the truck is included, the following formula is obtained:

$$k \leq \frac{m}{0.75}\frac{1-A_w}{A_w} \tag{7.8}$$

It is important to read this inequality in the proper way.

At first the information contained in it is: in order to fulfil the heavy traffic situation in the service station when the steady-state availability of machines in work is known and the number of these machines is m, the number of repair stands k should be at most equal to the right side of formula (7.8). Notice that here there is no information on an advisable number of repair stands for the machinery system.

It should also be realized that the parameter m, a component of formula (7.8) is constant here, i.e. the intensity of arrivals to the service system is stable and equals $m\delta$, but in this operational reality the number of machines in the work state is a random variable, at most equalling m. Furthermore, if the reserve is included in the considerations, it can easily be concluded that the large number of spares strengthens the stability of the arrivals intensity. Following this line of consideration, the conclusion is reached that the parameter k, the number of repair stands able to work, is also a random variable. The repair stand can also fail. The situation has become complicated. The following discussion tries to resolve this.

First, the problem of the unreliability of repair stands. Their intensity of failures is usually significantly lower than the intensity of failures of machines. Moreover, in mines there is great pressure to keep the workshop running, so the management does different things to ensure the continuous operation of such shops. In the following discussion it is therefore assumed that the intensity of failures of stands is negligible. However, in section 7.6 the problem of the reliability of repair stands will be considered comprehensively.

The second problem can be solved in the following way. Initially formula (7.8) was formulated assuming that the parameter equals m. Later, when the probability distribution of the number of failed machines is known, different system parameters can be calculated, among other things the expected value of the number of trucks in the work state E_p. If this value is known, then m can be replaced by E_p and checked again to see whether the appropriate conditions are fulfilled.

Bearing in mind the inequality (7.6), only those formulas from the set of formulas (2.2) to (2.9) that fulfil this inequality should be used. Therefore, the procedure of construction of the probability distribution of the number of machines in failure is as follows.

1. Power exponents' calculation:

$$\beta_1 = 2m\xi \frac{C_M + C_R}{C_R^{\,2}} \quad \beta_5 = 2(m+r)\xi \frac{C_M + C_R}{\left(C_R - \xi C_M\right)^2} \quad \beta_3 = \frac{2k}{\xi}\frac{C_M + C_R}{C_R^2}. \tag{7.9}$$

2. Construction of functions $g(x)$:

$$g_4(x) = \frac{1}{m\xi C_M + xC_R}\left(\frac{m\xi C_M + xC_R}{m\xi C_M}\right)^{\beta_1} \exp\left(\frac{-2x}{C_R}\right)$$

$$g_5(x) = \frac{1}{(m+r-x)\xi C_M + xC_R}\left(\frac{(m+r-x)\xi C_M + xC_R}{m\xi C_M + rC_R}\right)^{\beta_5} \exp\left(\frac{-2(\xi+1)(x-r)}{\xi C_M - C_R}\right)$$

$$\text{for } \kappa C_M - C_R \neq 0 \tag{7.10}$$

$$g_5(x) = \frac{1}{(m+r)\xi C_M}\exp\left(\frac{2(x-r)}{C_M} - \frac{(\xi+1)(x^2-r^2)}{(m+r)\xi C_M}\right) \quad \text{for } \kappa C_M - C_R = 0$$

$$g_6(x) = \frac{1}{(m+r-x)\xi C_M + kC_R}\left(\frac{(m+r-x)\xi C_M + kC_R}{(m+r-k)\xi C_M + kC_R}\right)^{\beta_3} \exp\left(\frac{2(x-k)}{C_M}\right).$$

3. Determination of the constants coming from the condition of function $h(x)$ continuity

$$\alpha_1 = \frac{g_6(k)}{g_5(k)} \quad \alpha_2 = \frac{g_5(r)}{g_4(r)}. \tag{7.11}$$

4. Computation of the constants K:

$$K_6 = \left(\alpha_1\alpha_2\int_0^r g_4(x)dx + \alpha_1\int_r^k g_5(x)dx + \int_k^{m+r} g_6(x)dx\right)^{-1}$$

$$K_5 = \alpha_1 K_6 \quad K_4 = \alpha_1\alpha_2 K_6. \tag{7.12}$$

5. Construction of the probability distribution of number of failed machines P_j:

$$\begin{aligned}
P_0 &= \int_0^{0.5} K_4 g_4(x)dx \\
P_j &= \int_{j-0.5}^{j+0.5} K_4 g_4(x)dx \quad \text{for } j = 1, 2, \ldots, r-1 \\
P_r &= \int_{r-0.5}^{r} K_4 g_4(x)dx + \int_r^{r+0.5} K_5 g_5(x)dx \\
P_j &= \int_{j-0.5}^{j+0.5} K_5 g_5(x)dx \quad \text{for } j = r+1, \ldots, k-1 \\
P_k &= \int_{k-0.5}^{k} K_5 g_5(x)dx + \int_k^{k+0.5} K_6 g_6(x)dx \\
P_j &= \int_{j-0.5}^{j+0.5} K_6 g_6(x)dx \quad \text{for } j = k+1, \ldots, m+r-1 \\
P_{m+r} &= \int_{m+r-0.5}^{m+r} K_6 g_6(x)dx.
\end{aligned} \tag{7.13}$$

The above probability distribution $P_j; j = 0,\ldots, m + r$ is very important. It allows for:

- Determination of several essential operation parameters of the truck-workshop system
- Detection of regularities that have an influence on it
- Verification of the quality of the selection of the system structural parameters: m, r and k.

Moreover, this distribution is a base for the construction of the probability distribution of the number of trucks in work state—information that is needed for further modelling and analysis.

The probability distribution of the number of failed trucks is a function of seven parameters:

$$P_j = f(m, r, k, \delta, \gamma, \sigma_p, \sigma_n)$$

where the last four are parameters that characterize the reliability properties of trucks and—to some extent—repair stands. The first three parameters are obtained during the procedure of selection of structural system parameters.

For proper analysis of this distribution it is necessary to keep in mind the conditions (2.10) and (2.11) that say firstly that the service system must be designed in such way that the intensity of service should be greater than the intensity of arrivals to the system, and secondly that the number of service stands should not be relatively great. The second stipulation is not dictated by the cost of construction and maintenance of these stands but because of the necessity of fulfilment of the heavy traffic condition. This condition is associated with the goodness of assessment of the system parameters. For further considerations, it is assumed that this condition is fulfilled as far as the operation of the repair shop is concerned.

The following is an analysis of the system.

It considers two-example systems[6] of different reliability of trucks; the remaining parameters will be the same[7]:

$$\mathfrak{S}_{\mathrm{I}} : < m = 50, r = 10, k = 16; \delta = 0.030; \sigma_p = 33; \gamma = 0.125; \sigma_n = 5 >$$
$$\mathfrak{S}_{\mathrm{II}} : < m = 50, r = 10, k = 16; \delta = 0.039; \sigma_p = 26; \gamma = 0.122; \sigma_n = 4 >.$$

Looking at the parameters[8] of these systems it is easy to deduce that the steady-state availability of trucks in the first system is 0.806, and in the second system is 0.758. The times of work state are following the exponential low, therefore, the mean equals the standard deviation. This rule does not hold in the state of repair.

Before beginning an analysis of these systems, it is necessary to check whether the appropriate conditions are fulfilled. Taking into account formulas (7.7) and (7.8) and remembering inequality (2.11), the following inequality is obtained for the steady-state availability of trucks:

$$\frac{m}{m+k} < A_w < \frac{4m}{4m+3k} \tag{7.14}$$

which gives for the above systems:

$$0.758 < A_w < 0.806.$$

This result means that the steady-state availability of trucks just fits the boundaries. At this point, the question arises: *what will happen if the trucks are of lower or higher steady-state availability than that determined by these limited values*?

---

[6] Systems will be marked by the gothic letter $\mathfrak{S}$. Mathematicians involved in considerations in theory of sets have assumed that sets will be marked just by gothic letters. Also relations and states will be marked here by gothic letters.

[7] The notation: $\mathfrak{S} : < x, y >$ means: system is determined by x and y.

[8] All parameters having a time unit are presumed h or h^{-1} if not specified in a different way.

If trucks are of poor quality, of lower availability ($A_w < 0.758$), then the intensity of arrivals to the shop will be greater than the shop's possibilities of repairing. An almost permanent queue of trucks is certain.

If trucks are of good, high availability ($A_w > 0.806$), then the intensity of arrivals to the shop will not be so demanding (good for repair people), not satisfying the heavy traffic situation.

The probability distributions for the number of failed trucks for the above systems are shown in Figure 7.8.

Looking at Figure 7.8 it is easy to observe that there are visible changes in the mass of probability. For trucks that are more reliable the mass of probability has low dispersion and is closer to the inception point of the horizontal axis. This means that on average there are a smaller number of machines in failure with a smaller standard deviation in this system. For trucks of lower reliability the mass of probability is flatter, with higher dispersion, the mean number of trucks in failure is greater. Taking into account the fact that the number of repair stands is 14, it can be predicted that a truck system of lower reliability will have problems being served, with so many machines in failure.

If the exploitation parameters of these two truck systems are now gathered. To make the analysis more comprehensive, a few more interesting measures[9] will be defined.

It is a well-known fact that a queue of clients can appear before each service system. Taking into account the reality of the mine, it is certain that this type of situation will take place. Analyzing formulas (2.5), a formula can be found to determine the mean time of truck waiting for repair if it stays at the shop. It is given by:

$$T_{ow} = \gamma^{-1} \sum_{j=k+1}^{m+r} (j-k) \int_{j-0.5}^{j+0.5} K_6 g_6(x)dx. \tag{7.15}$$

This parameter is conditional. On average, a truck waits for repair T_{ow} provided that the shop is full. The failed truck is moved to the shop if there is a repair stand free.

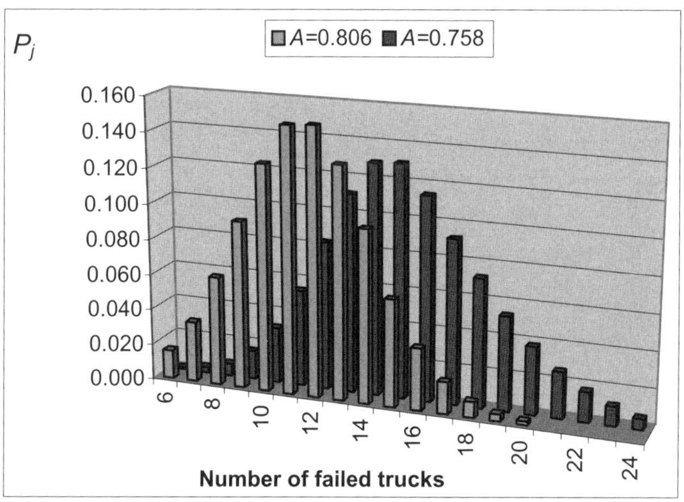

Figure 7.8. Probability distributions P_j for the numbers of failed trucks for different steady-state availability $A_w = 0.806$ and 0.758 for the system of structural parameters $<m = 50, r = 10, k = 16>$.

[9]It is important to realize that all the interesting parameters of the system that are achieved from the distribution (7.13) are the approximate ones only because the whole modus operandi only has an approximate character.

This parameter is important and must be taken into consideration in later stages of modelling. Note that here a new truck state has just occurred—a state in which the truck waits for repair, forming a queue. If the realization of the transportation task is taken into consideration, then the conclusion is that this state absorbs trucks for a certain period of time, and the mean time of this absorption is just T_{ow}. For the accomplishment of the transportation task, it makes no difference whether the truck waits for repair or is being repaired. It is incapable of work. Following this mode of consideration, these two states can be lumped into one—*state of unserviceability of truck*. The mean time of this state T_{ns} is the sum of two components: the mean time of truck repair plus the unconditional mean time of the truck spent in a queue waiting for repair. Analyzing formulas (2.5) and (7.15) gives:

$$T_{ns} = T_n (1 + \phi) \text{ where: } \phi = \int_{k+0.5}^{m+r} K_6 g_6(x)dx \sum_{j=k+1}^{m+r} (j-k) \int_{j-0.5}^{j+0.5} K_6 g_6(x)dx. \tag{7.16}$$

At this point, the two previously defined parameters need to be reviewed and some corrections made.

First, the steady-state availability of the truck needs to be modified. Because it is the probability of an event that at any moment the truck is in the work state, a new measure of availability has to be formulated, that is an adjustment of the original one. This measure is referred to as the *adjusted steady-state availability*. It is determined by the formula:

$$A_w = \frac{\delta^{-1}}{\delta^{-1} + T_{ns}} \qquad A'_w \leq A_w. \tag{7.17}$$

A second parameter that has to be modified is the *adjusted flow intensity rate*, marked by ρ'. If it is stated that γ needs to be replaced by T_{ns}^{-1}, then consequently the flow intensity rate needs to be taken into account, because this rate is important for assessing the serviceability of the shop. Hence the following formula is generated:

$$\rho' = \frac{E_p \delta}{k T_{ns}^{-1}} \tag{7.18}$$

and this is a definition of the adjusted flow intensity rate.

The following presents all the parameters of these two truck systems that have a back-up facility—the repair shop of known parameters. The values of these parameters are given in Table 7.1. This table also contains the mean number of failed machines E_u and the mean number of trucks in work state E_p that can be calculated from the classical definition of the expected value[10] for a discrete type of random variable:

$$E(\boldsymbol{X}) = \sum_i x_i p(x_i)$$

but the functions $g(x)$ can be used.

Hence, the expected number of machines in failure can be computed from the formula:

$$E(\boldsymbol{X}) = E_u = K_4 \int_0^r x g_4(x)dx + K_5 \int_r^k x g_5(x)dx + K_6 \int_k^{m+r} x g_6(x)dx. \tag{7.19}$$

[10] A definition of the expected value of a random variable of continuous type is given by formula (5.1).

Table 7.1. Parameters of example systems $\mathfrak{S}_I$ and $\mathfrak{S}_{II}$.

System	Flow intensity rate	Mean time of work state of truck	Mean time of repair of truck	Steady-state availability of truck	Mean number of failed trucks	Mean number of trucks in work state	Mean number of trucks in reserve	Conditional mean time of truck waiting for repair	Mean time of truck unserviceability	Adjusted steady-state availability of truck	Adjusted flow intensity rate	Mean number of busy repair stands	Mean number of idle repair stands	Operating utilization of system
	ρ	δ^{-1}	γ^{1}	A_w	E_u	E_p	E_r	T_{ow}	T_{ns}	A'_w	ρ'	Ek	EI	OU
		h	h		trucks	trucks	trucks	h	h			stands	stands	
$\mathfrak{S}_I$	0.750	33	8	0.806	11.6	48	0.4	0.5	8.0	0.806	0.72	11.5	4.5	0.72
$\mathfrak{S}_{II}$	0.999	26	8.2	0.758	15.3	44.7	0.0	8	11.2	0.696	1.22	14.3	1.7	0.89

In the same way, the expected number of machines in work state is given by:

$$E_p = m - \int_r^k (x-r)K_5 g_5(x)dx - \int_k^{m+r} (x-r)K_6 g_6(x)dx. \tag{7.20}$$

Certainly, an interesting system parameter is the expected number of busy repair stands, given by:

$$E_{BRS} = k - \int_0^r (k-r)K_4 g_4(x)dx - \int_r^k (k-r)K_5 g_5(x)dx \tag{7.21}$$

as well as the expected number of idle repair stands, determined by the formula:

$$E_{IRS} = \int_0^r (k-x)K_4 g_4(x)dx + \int_r^k (k-x)K_5 g_5(x)dx = k - E_{BRS}. \tag{7.22}$$

The operative utilization of the repair shop is given by the ratio:

$$OU = E_{BRS}/k. \tag{7.23}$$

The operative utilization represents the fraction of busy repair stands and is a measure of the efficiency of the repair shop.

Finally, notice that having a total of $m + r$ trucks in the system, and knowing that the mean number of failed trucks is E_u and the mean number of trucks in work state is E_p, the mean number of trucks in reserve is calculated as:

$$E_r = m + r - E_u - E_p. \tag{7.24}$$

The figures contained in the table can now be analyzed.

The difference between the steady-state availability of trucks does not appear great, with only 6% lower availability of machines in the second system. But changes in some parameters are very significant. This change in availability made an increase in the number of failed machines by 32% and a decrease of the number of trucks in the work state by 7%. The conditional mean time of the truck waiting for repair increased 16 times, which increased the mean time of truck unserviceability by 40%. Due to these reasons the 'real' (adjusted) steady-state availability dropped. The adjusted flow intensity rate exceeded unity by 22%, meaning that an almost permanent queue is observed. In such a situation, an increase in the number of repair stands seems highly advisable. The mean number of busy repair stands increased by 24% and the expected value of the number of idle repair stands decreased almost 3 times. The operative utilization of the repair shop increased by 24%.

The first system looks well designed: 48 of 50 trucks are in the work state on average, i.e. 96%. Queues of trucks waiting for repair seldom occur—so rarely that there is no significant increase in the time of truck unserviceability. Principally, the state of unserviceability consists almost exclusively of the repair state. The adjusted steady-state availability remained intact, whereas the flow intensity rate dropped slightly.

In summarizing, one conclusion is irresistible—small changes in truck availability made great changes in the parameters of their system with the repair shop. Taking into account the financial side of truck operation, small changes in truck reliability can make considerable changes in the profitability of machinery system operation/exploitation.

The next parameter to consider is truck reserve.

The following is a comparison of the probability distributions of a number of trucks in failure for two identical machinery systems with a different number of spare machines, e.g.:

$$\mathfrak{S}_{\text{III}} : < m = 60,\ r = 8,\ k = 27;\ A_w = 0.709 >$$

and

$$\mathfrak{S}_{\text{IV}} : < m = 60,\ r = 16,\ k = 27;\ A_w = 0.709 >.$$

The probability distributions for the number of failed trucks for the above systems are shown in Figure 7.9.

Looking at Figure 7.9 it is hard to observe significant changes in the probability except for the fact that the number of spare trucks has been doubled.

The problem of truck reserve can be approached in a different way.
Consider two systems:

$$\mathfrak{S}_{\text{V}} : < m = 53,\ r,\ k = 20;\ A_w = 0.728,\ \sigma_p = 24;\ \rho = 0.99 >$$

and

$$\mathfrak{S}_{\text{VI}} : < m = 53,\ r,\ k = 20;\ A_w = 0.781,\ \sigma_p = 30;\ \rho = 0.74 >$$

for $r = 6, 8, 10, 12, 14$.

Similarly to the first two example systems $\mathfrak{S}_{\text{I}}$ and $\mathfrak{S}_{\text{II}}$, the system parameters have been calculated and the results are given in Table 7.2.

Even at first glance, the values of some parameters and character of some relationships are quite interesting. These will be considered in sequence.

Figure 7.10 shows the relationship between the truck mean time of the state of unserviceability T_{ns} and the truck reserve size r, whereas Figure 7.11 presents the relationship between the conditional mean time of truck waiting for repair T_{ns} and the truck reserve size r.

The above two figures give the same information—by applying trucks of lower availability there is a significant increase in both the conditional mean time of the truck waiting for repair at the repair shop T_{ns} and the truck mean time of state of unserviceability T_{ow}. This is the reason that the adjusted steady-state availability of trucks A' dropped—Figure 7.12.

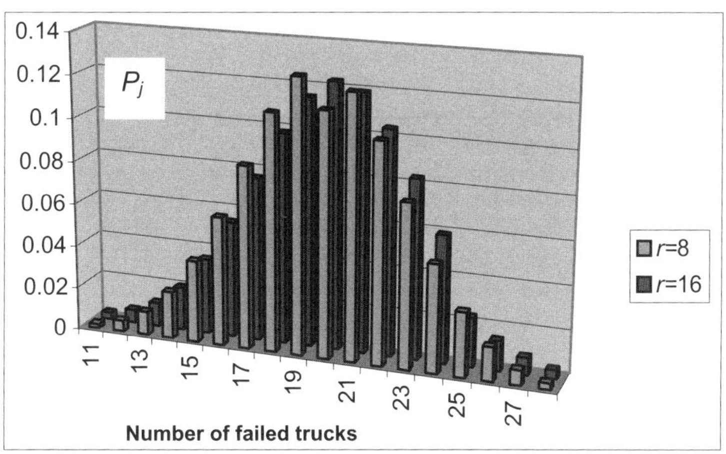

Figure 7.9. Probability distributions P_j of numbers of failed trucks for different reserve size $r = 8$ and $r = 16$ for the system of $m = 60$ trucks and $k = 27$ repair stands.

Table 7.2. Parameters of example systems $\mathfrak{S}_V$ and $\mathfrak{S}_{VI}$.

System	Input parameters	unit	r	6	8	10	12	14
$\mathfrak{S}_V$	$m = 53$, $k = 20$, $A = 0.781$, $\rho = 0.74$	h	T_{ow}	0.1	0.2	0.3	0.5	0.7
		h	T_{ns}	8.5	8.5	8.5	8.5	8.5
		trucks	E_u	12.9	13.3	13.8	14.1	14.5
		trucks	E_p	46.1	47.6	49.1	50.4	51.4
		trucks	E_r	0	0	0.1	0.5	1.1
			A'	0.781	0.781	0.781	0.781	0.781
			ρ'	0.64	0.67	0.69	0.71	0.72
		stands	E_k	12.9	13.3	13.7	14.1	14.4
		stands	E_l	7.1	6.7	6.3	5.9	5.6
			OU	0.64	0.67	0.69	0.71	0.72
System	**Input parameters**		r	6	8	10	12	14
$\mathfrak{S}_{VI}$	$m = 53$, $k = 20$, $A = 0.728$, $\rho = 0.99$	h	T_{ow}	1.9	2.3	4.7	7.0	10.3
		h	T_{ns}	9.3	9.6	10.1	11.1	12.8
		trucks	E_u	16.2	16.8	17.5	18.2	19.0
		trucks	E_p	42.8	44.2	45.5	46.8	47.9
		trucks	E_r	0	0	0	0	0.1
			A'	0.724	0.718	0.707	0.687	0.655
			ρ'	0.82	0.87	0.94	1.07	1.26
		stands	E_k	16	16.5	17	17.4	17.8
		stands	E_l	4	3.5	3	2.6	2.1
			OU	0.80	0.82	0.85	0.87	0.89

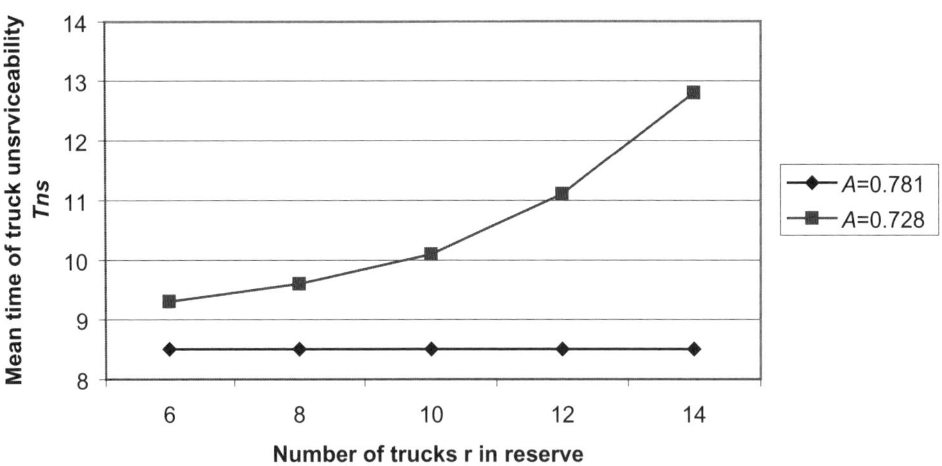

Figure 7.10. Truck mean time of state of unserviceability T_{ns} versus truck reserve size r for system $\mathfrak{S}_V$ and $\mathfrak{S}_{VI}$.

If this information is translated into a more practical form. By purchasing machines of medium reliability, a lot of work is created to keep these machines going. The only solution here to improve the situation would be enlargement of the back-up facility. Therefore, what has been saved by purchasing such machines is lost by increasing the number of repair stands. More diagnostic and repair equipment is needed, more people to mend and maintain, more spare parts, etc. In some cases, it will be a problem to arrange additional room to enlarge the existing shop or build a new one.

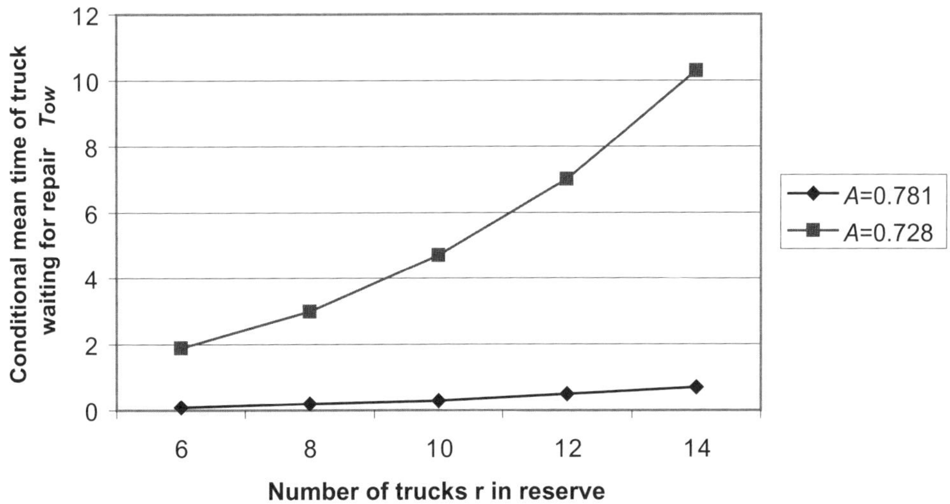

Figure 7.11. Conditional mean time of truck waiting for repair T_{ow} versus truck reserve size r for system $\mathfrak{S}_{V}$ and $\mathfrak{S}_{VI}$.

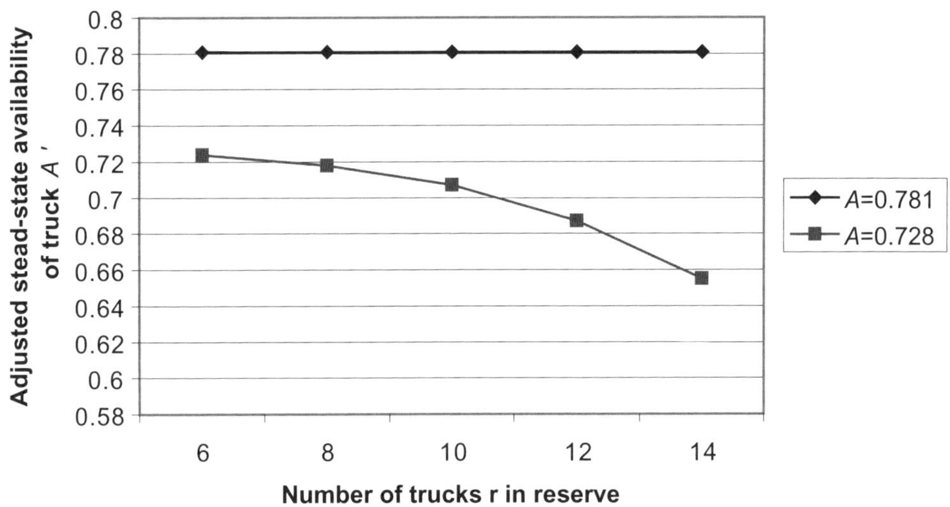

Figure 7.12. Adjusted steady-state availability of trucks A' versus truck reserve size r for system $\mathfrak{S}_{V}$ and $\mathfrak{S}_{VI}$.

It is worth pointing out here that in some technical publications concerning technical systems it is suggested that increasing the number of spare units for a given system generally increases its performance and—speculatively—its profitability. Looking at the above figures leads to doubt as to whether this is in general true.

An increase is visible in the number of trucks in the work state (Figure 7.14), but whether the profit obtained from it will cover the necessary expenses depends on the particular case. An additional problem is flow rate value. This will be discussed on the next page.

The following discussion concerns the four relationships jointly. These are presented in:

- Figure 7.13: the mean number of trucks in failure E_u versus the number of trucks r in reserve
- Figure 7.14: the mean number of trucks in work state E_p versus the reserve size

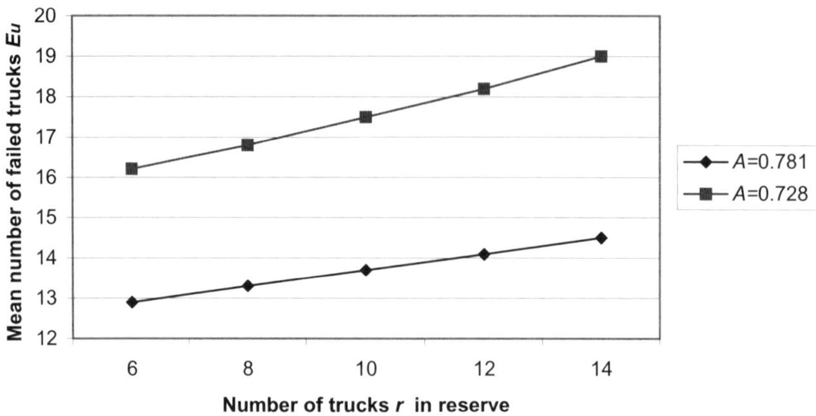

Figure 7.13. Mean number of failed trucks E_u versus truck reserve size r for system $\mathfrak{S}_{V}$ and $\mathfrak{S}_{VI}$.

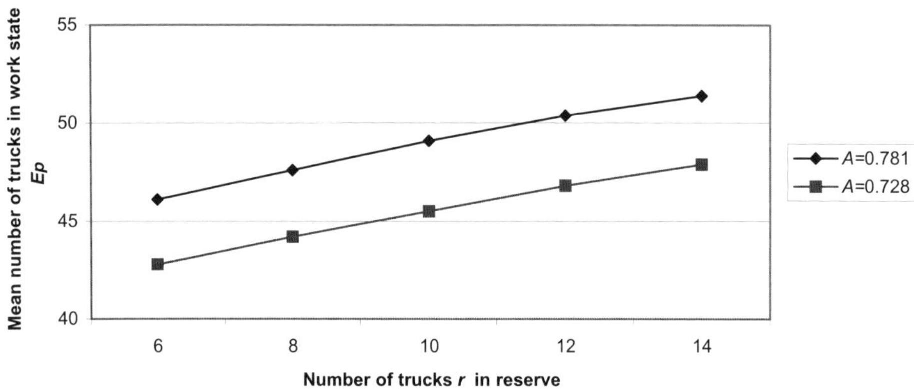

Figure 7.14. Mean number of trucks in work state E_p versus truck reserve size r for system $\mathfrak{S}_{V}$ and $\mathfrak{S}_{VI}$.

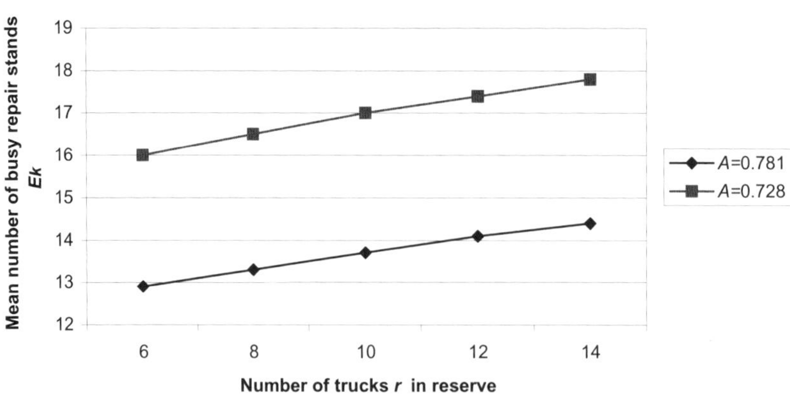

Figure 7.15. Mean number of busy repair stands E_k versus truck reserve size r for system $\mathfrak{S}_{V}$ and $\mathfrak{S}_{VI}$.

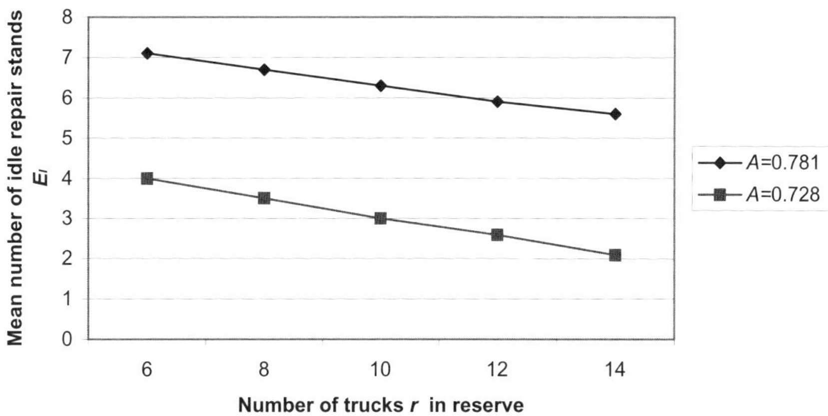

Figure 7.16. Mean number of idle repair stands E_k versus truck reserve size r for system $\mathfrak{S}_{V}$ and $\mathfrak{S}_{VI}$.

- Figure 7.15: the mean number of busy repair stands E_k versus the reserve size
- Figure 7.16 the mean number of idle repair stands E_l versus the reserve size.

For both levels of availability a similar increment is observed—an almost strict linear relationship between these parameters. The increment/decrement can be assumed to be constant. The formula describing a straight line possesses two parameters: one determining the angle of inclination of the line, the second where the line crosses the axis. In this case, both these parameters are determined by the properties of the machinery system given. It makes no sense to compare lines associated with different systems.

The straight-line regularity can be noticed applying different system parameters, but these considerations have so far been confined to the theoretical scope of concern. However, if is important to consider mine reality. In a pit a certain number of trucks flows between loading machines and dumping points. But the area is limited, haulage roads are of a determined capacity, and working areas around shovels are sometimes confined. All these factors indicate that the number of hauling units cannot be increased in a completely free way.

Notice the following regularity. When there are a small number of trucks in the pit, they flow with barely any problem. Increasing the number of transporting machines will at first not be a problem. But when the number of machines is relatively large, they start to disturb each other. Sometimes queues will appear not only in front of shovels but also in different parts of the pit. The flow rate of such a system drops.

Figure 7.17 gives different information. With transporting machines of low reliability, it is necessary to be aware that an increasing number of circulating machines causes growth in the intensity flow rate. The limited value is 1. Hence, if an incorrect decision is made and there are many spare machines, directing them all to operate will be the wrong decision. It will choke, causing queues. In this situation, it is better to keep some machines in a cold reserve.

The next parameter to be taken into consideration is the number of repair stands k. This is an extremely important parameter that decides—above all—the number of trucks in work state and the size of the repair shop.

As a preliminary consideration, here are some remarks referring to mine reality. A stream of machines is flowing into the shop. The character of this stream is non-homogeneous. A machine is directed to the shop because a failure has occurred so a repair is required, or the machine needs some maintenance that can be of a different nature—some operation materials need to be changed, the parameters of some assemblies need adjusting, a general technical survey is required, etc. The scope of consideration of this monograph is exclusively stochastic phenomena occurring during exploitation/operation process of machines in the system. The determined schedules of surveys of

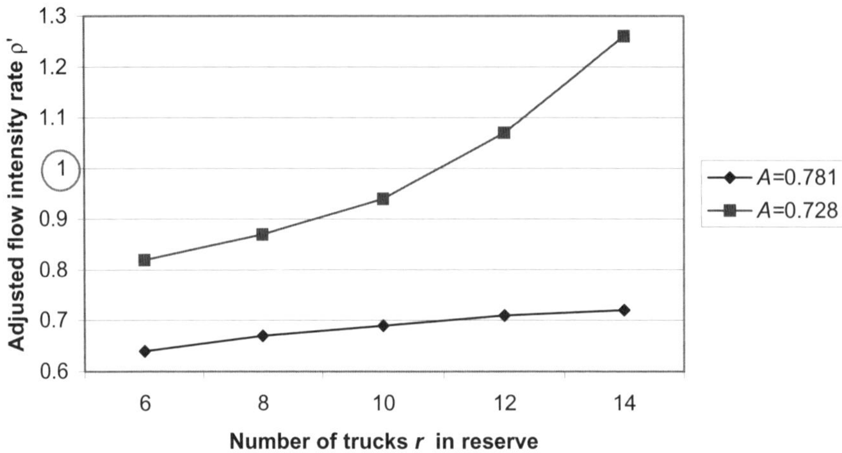

Figure 7.17. Adjusted flow intensity rate ρ' versus truck reserve size r for system $\mathfrak{S}_V$ and $\mathfrak{S}_{VI}$.

machines are not considered. For this reason, when considerations are directed at the selection of structural parameters the optimization of the number of repair stands will not be given.

Consider the system parameters of these two systems;

$$\mathfrak{S}_{VII} : < m = 66, r = 13, k; A_w = 0.753, \delta = 0.040, \gamma = 0.122 >$$

$$\text{for } k = 21, 22, 23, 24, 25, 26, 27, 28, 30$$

and

$$\mathfrak{S}_{VIII} : < m = 66, r = 13, k; A_w = 0.811, \delta = 0.030, \gamma = 0.129 >$$

$$\text{for } k = 16, 17, 18, 19, 20, 21, 22, 23$$

repair stands.

Table 7.3 contains results of the computations of the following:

- Flow rate intensity ρ
- Conditional mean time of the truck awaiting repair T_{ow}
- Mean time of truck unserviceability T_{ns}
- Adjusted steady-state availability of truck A'
- Mean number of failed trucks E_u
- Standard deviation of number of failed trucks D_u
- Mean number of trucks in work state E_p
- Adjusted flow rate intensity ρ'
- Mean number of busy repair stands E_k
- Mean number of idle repair stands E_I
- Operative utilization *OU*.

All these parameters have been determined except the standard deviation D_u. The definition of this statistics parameter was described by formula (5.2). This measure of values of random variable dispersion will be needed to support inference on the behaviour of this random variable. In Table 7.3, two auxiliary probabilistic measures are included, namely:

Table 7.3. Parameters of example systems $\mathfrak{S}_{VII}$ and $\mathfrak{S}_{VIII}$.

System	Input parameters	unit	k	16	17	18	19	20	21	22
			ρ	0.96	0.90	0.85	0.81	0.77	0.73	0.70
		h	T_{ow}	16.7	9.0	4.7	2.4	1.2	0.6	0.3
		h	T_{ns}	16.1	10.8	8.8	8.1	7.8	7.8	7.8
			A'	0.675	0.755	0.791	0.805	0.810	0.811	0.811
		trucks	E_u	16.57	15.75	15.29	15.05	14.91	14.85	14.82
	$m = 66$	trucks	D_u	4.78	4.26	3.9	3.67	3.53	3.46	3.42
$\mathfrak{S}_{VII}$	$r = 13$	trucks	E_p	61.9	62.7	63.1	63.3	63.5	63.8	63.8
	$A = 0.811$		ρ'	1.86	1.20	0.92	0.81	0.75	0.71	0.67
		stands	E_k	14.41	14.58	14.69	14.73	14.76	14.77	14.78
		stands	E_I	1.6	2.4	3.3	4.3	5.2	6.2	8.2
			OU	0.9	0.86	0.82	0.78	0.74	0.7	0.67
		trucks	E_c	20.4	20.4	20.8	21.4	22	22.8	23.6
			P_{con}	0.496	0.341	0.219	0.133	0.076	0.041	0.021

System	Input parameters	unit	k	21	22	23	24	25	26	27	30
			ρ	1.03	0.98	0.94	0.90	0.87	0.83	0.8	0.72
		h	T_{ow}	14.3	8.2	4.5	2.4	1.3	0.6	0.3	0.0
		h	T_{ns}	14.5	10.7	9.1	8.5	8.3	8.2	8.2	8.2
			A'	0.675	0.705	0.733	0.746	0.751	0.752	0.753	0.753
		trucks	E_u	20.82	20.25	19.91	19.72	19.61	19.55	19.52	19.51
	$m = 66$	trucks	D_u	4.64	4.27	4.01	3.83	3.72	3.65	3.62	3.58
$\mathfrak{S}_{VIII}$	$k = 13$	trucks	E_p	58.1	58.7	59.0	59.2	59.3	59.4	59.4	59.5
	$A = 0.753$		ρ'	1.61	1.14	0.94	0.84	0.79	0.75	0.72	0.65
		stands	E_k	19.06	19.25	19.36	19.42	19.46	19.48	19.49	19.49
		stands	E_I	1.9	2.8	3.6	4.6	5.5	6.5	7.5	10.5
			OU	0.91	0.87	0.84	0.81	0.78	0.75	0.72	0.65
		trucks	E_c	25	25.3	25.7	26.4	27.1	27.8	28.6	31.3
			P_{con}	0.441	0.308	0.203	0.127	0.075	0.042	0.022	0.002

- The conditional mean number of trucks waiting for repair at the shop provided that the shop is fully occupied (busy) $E_c = E(X|X > k)$; this is calculated from the classical formula for the conditional mean, i.e.

$$E_c = E(\boldsymbol{X}) \mid \boldsymbol{X} > K) = \frac{\int_k^{m+r} x g_6(x)dx}{\int_k^{m+r} g_6(x)dx} \tag{7.25}$$

- The probability of such a situation happening, i.e. the number of failed trucks exceeds the number of repair stands in the shop i.e. $P(\boldsymbol{X} > k)$

$$P_{con} = \int_{k}^{m+r} K_6 g_6(x)dx \tag{7.26}$$

The following is an analysis of the information contained in this table.

In the first row values of the flow rate intensity ρ calculated based on the opening information are roughly correct but—after calculation of the system parameters—some values of the adjusted flow rate intensity significantly exceeded boundaries.

A short conclusion here seems appropriate:

if the value of the flow intensity rate is near the limit it is highly recommended to verify it by calculating its adjusted flow intensity rate.

A second obvious conclusion can be drawn by looking at the values of the conditional mean time of a truck waiting for repair T_{ow} and the mean time of truck unserviceability T_{ns}:

increasing the number of repair stands k made a decrease in the truck queue at the shop and the adjusted steady-state availability of the truck tends towards the limited value A.

Figures 7.18, 7.19 and 7.20 show example plots of these relationships.

If a queue of failed trucks waiting for repair is considered, it is interesting how many of these units will be waiting at the shop. The answer to this question gives values of the E_c parameter—the conditional mean value of the number of trucks waiting for repair at the shop if the shop is fully occupied. The values given in the table may be terrifying at first glance, but it is necessary to remember that they are conditional ones. If the values of the last row are considered, the situation is cooled down—some events are rather rare. If the probability is high and the number of trucks at the shop is significant there is no doubt that an increase in number of repair stands is required.

The next relationship is extremely important: the mean number of trucks in failure E_u against the number of repair stands. At this point, it is worth recalling that the Maryanovitch system possesses one very advantageous property: there is no queue at the shop, i.e. no truck time is lost owing to this. When looking at a large machinery system a key question arises: *is it possible to build a repair shop of a lower number of stands than the total number of machines in the system and to obtain a negligible queue of machines waiting for repair?*

At this point an answer can be obtained to this vital question, by considering this relationship.

Figure 7.21 shows the mean number of trucks in failure E_u against the number of repair stands k for trucks of the steady-state availability $A = 0.811$, and Figure 7.22 illustrates this relationship for trucks of the steady-state availability $A = 0.753$. Both figures give the same conclusion: the mean number of trucks in failure E_u against the number of repair stands k runs quickly towards a certain limited value. The number of stands for which the function is stabilized is far from the total number of machines—that is 79 for the system considered. Czaplicki (2006) presented this relationship, concluding that: *there is no need to build many repair stands to obtain the machinery system characteristics determined by the Maryanovitch model.*

The limited value E_u depends on the steady-state availability of the truck, and for practical purposes it can be assumed that for trucks of high reliability the stabilized value of E_u is reached when

$$k \cong (m + r)/2.5$$

whereas for trucks of medium reliability

$$k \cong (m + r)/3.5.$$

This argument can be supported by looking at the course of standard deviation D_u of this random variable. Figures 7.23 and 7.24 present these relationships.

Looking at the above figures, it can be concluded that the standard deviation D_u is also stabilized at approximately the same number of repair stands as for the corresponding mean value.

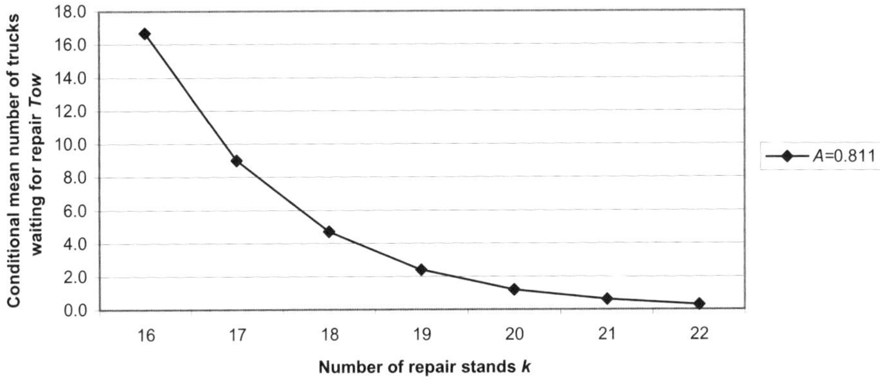

Figure 7.18. Conditional mean number of trucks waiting for repair T_{ow} versus number of repair stands k for trucks of steady-state availability $A = 0.811$.

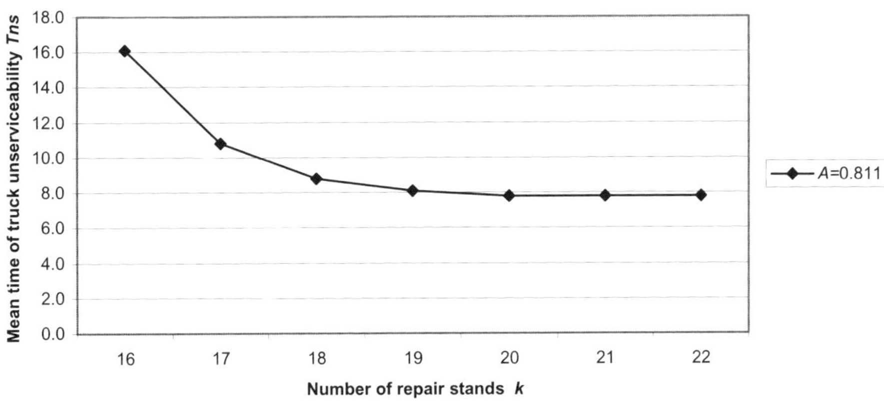

Figure 7.19. Mean time of truck unserviceability T_{ns} versus number of repair stands k for trucks of steady-state availability $A = 0.811$.

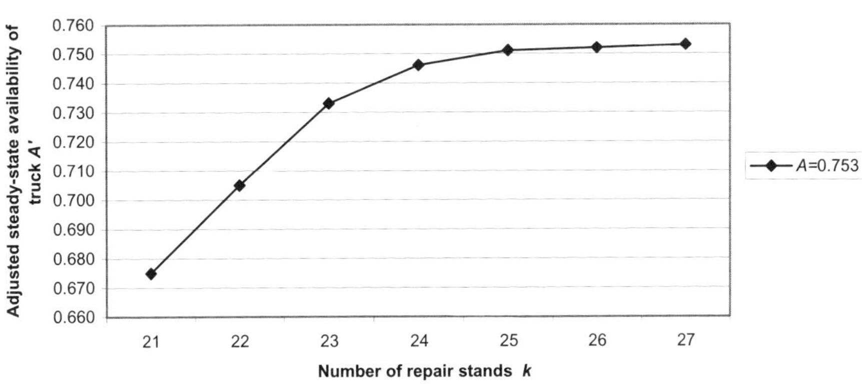

Figure 7.20. Adjusted steady-state availability of truck A' versus number of repair stands k for trucks of steady-state availability $A = 0.753$.

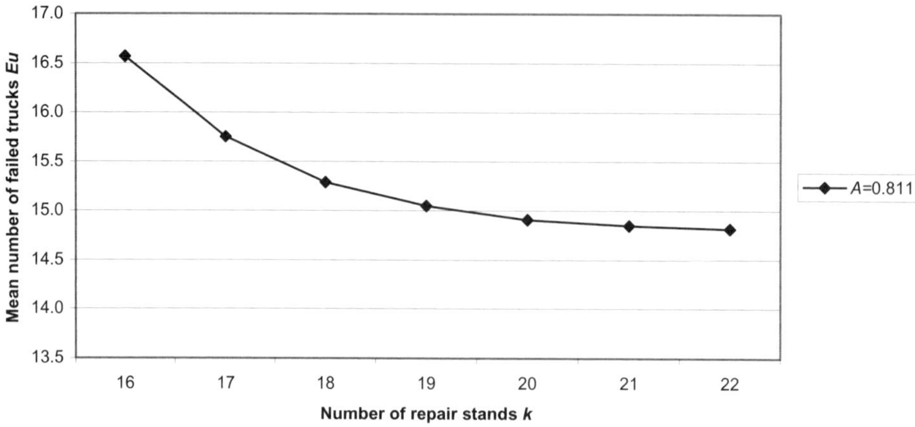

Figure 7.21. Mean number of trucks in failure E_u versus number of repair stands k for trucks of the steady-state availability $A = 0.811$.

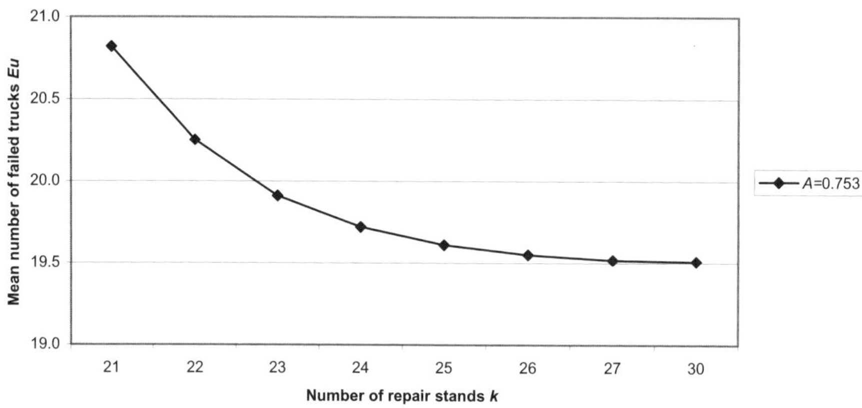

Figure 7.22. Mean number of trucks in failure E_u versus number of repair stands k for trucks of the steady-state availability $A = 0.753$.

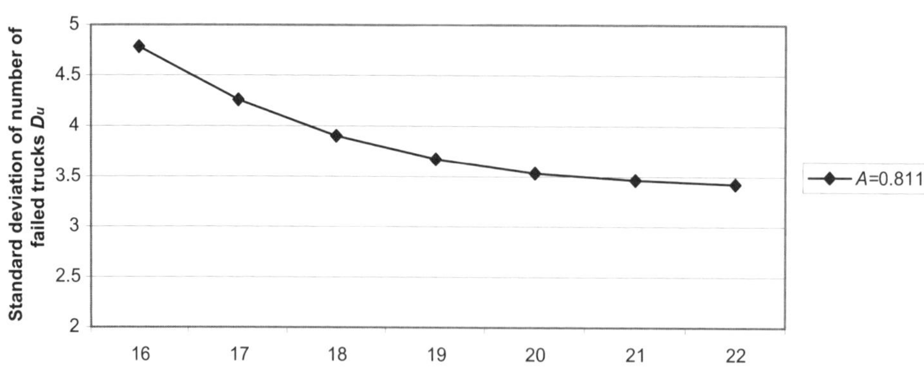

Figure 7.23. Standard deviation of number of trucks in failure D_u versus number of repair stands k for trucks of the steady-state availability $A = 0.811$.

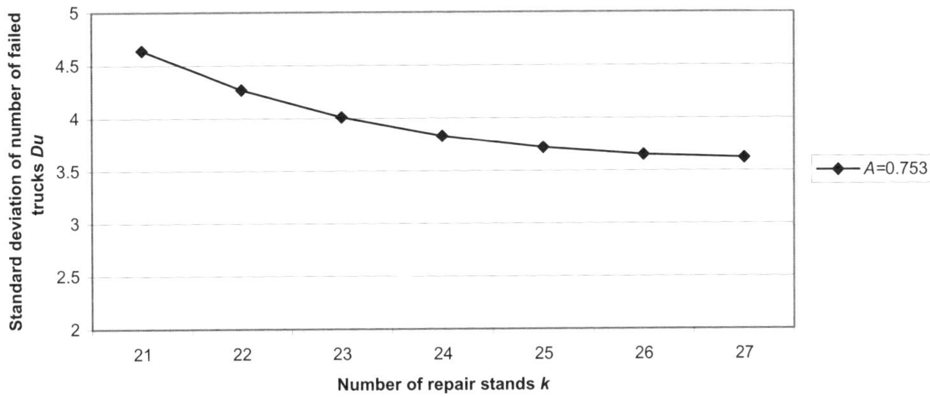

Figure 7.24. Standard deviation of number of trucks in failure D_u versus number of repair stands k for trucks of the steady-state availability $A = 0.753$.

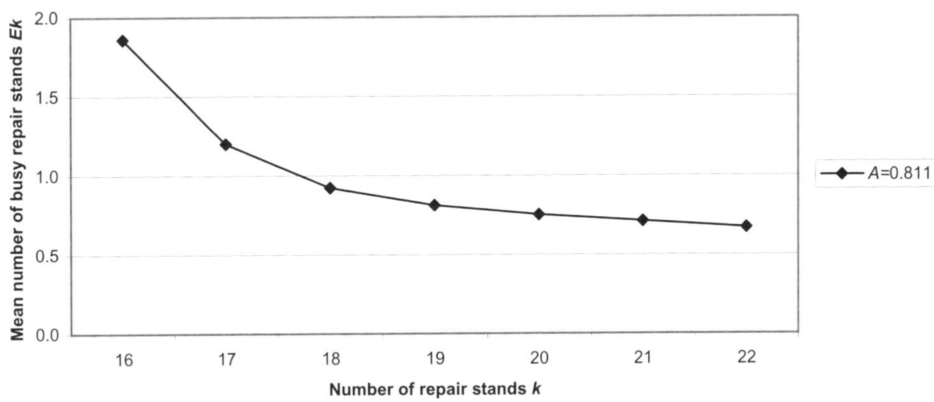

Figure 7.25. Mean number of busy repair stands E_k versus number of repair stands k for trucks of the steady-state availability $A = 0.811$.

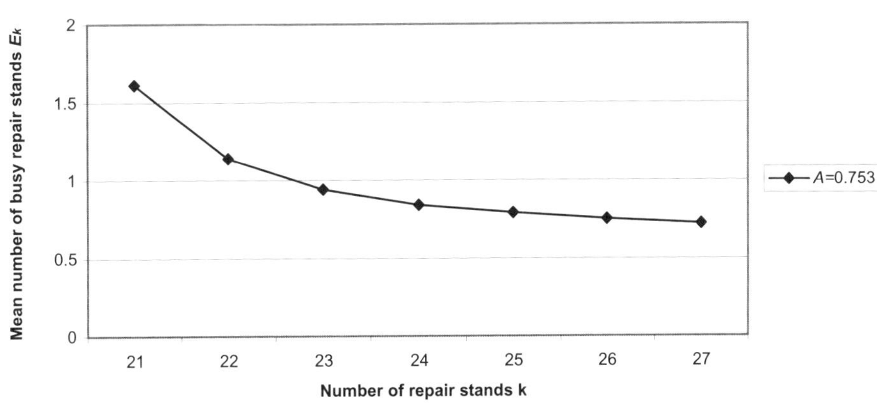

Figure 7.26. Mean number of busy repair stands E_k versus number of repair stands k for trucks of steady-state availability $A = 0.753$.

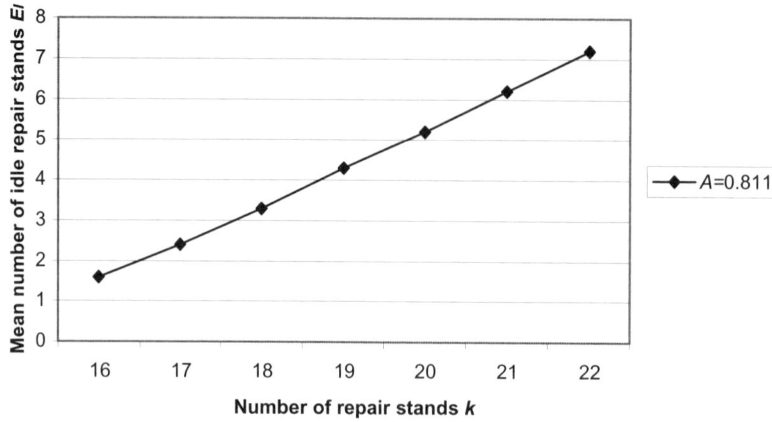

Figure 7.27. Mean number of idle repair stands E_l versus number of repair stands k for trucks of the steady-state availability $A = 0.811$.

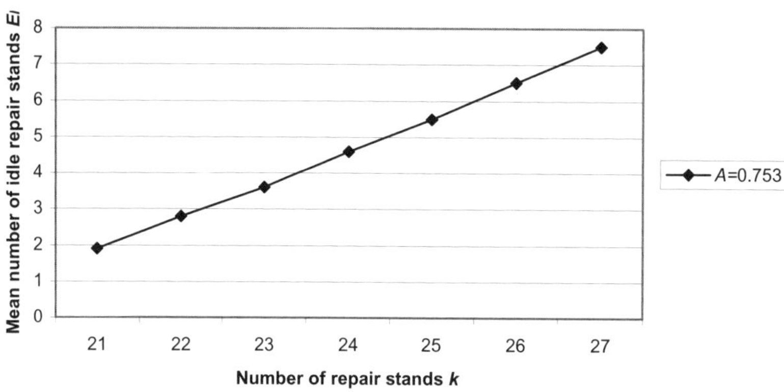

Figure 7.28. Mean number of idle repair stands E_l versus number of repair stands k for trucks of the steady-state availability $A = 0.753$.

Further evidence for this can be gleaned by considering the relationship between the mean number of busy repair stands E_k and the number of repair stands applied in the system. Figures 7.25 and 7.26 show this relationship for trucks of availability $A = 0.811$ and $A = 0.753$ respectively.

This important system regularity can now be named. If the number of repair stands in the machinery system is such that a further increase in the number of stands does not make a significant reduction in the number of failed machines, then a system with this number of repair stands will be called *a system with service saturation*.

The last relationship to be considered is the mean number of idle repair stands E_l against the number of repair stands in the shop. Figures 7.27 and 7.28 illustrate this relation.

It is interesting to note that this relationship is a linear one.

7.3 PROBABILITY DISTRIBUTION OF NUMBER OF TRUCKS IN WORK STATE

If the probability distribution of a number of failed trucks (7.13) is known, then the probability distribution of a number of trucks in work state can be determined. If $P_{wm}^{(p)}$ denotes the probability of an event that m trucks are in work state.

Looking at the set of patterns (7.13) the following can be written:

$$
\begin{cases}
P_{wm}^{(p)} = \sum_{j=0}^{r} P_j \\
P_{w(m-j)}^{(p)} = P_{r+j} \quad \text{for } j = 1, 2, \ldots, m
\end{cases}
\tag{7.27}
$$

where P_j is the probability that j trucks are in failure.

A similar probability distribution can be obtained from the Maryanovitch model (2.1).

Both probability distributions have the same regularity: accumulation of the mass of probability in the point $P_{wm}^{(p)}$. The reason for this is the constant number of working trucks when the number of failed machines varies from 0 to r. If the number of trucks in failure exceeds the reserve size the number of trucks in work state is less than m. Changes in the mass of probability are smooth, but have stepped characters because of the discrete nature of the random variable considered.

Figure 7.29 presents the probability distribution of a number of trucks in work state for the example systems:

$$\mathfrak{S}_{IX} : < m = 50, r = 10, k = 14; A_w = 0.763 >$$

and for the corresponding Maryanovitch model:

$$\mathfrak{S}_{X} : < m = 50, r = 10, k = m + r, A_w = 0.763 >.$$

For a system of trucks of medium reliability, changes in the allocation of the mass of probability are clearly visible. Remember that these changes are connected with the number of repair stands applied exclusively.

If the same probability distributions for the same machinery system but for trucks of high reliability, say $A_w = 0.811$ are considered, then Figure 7.30 shows these distributions.

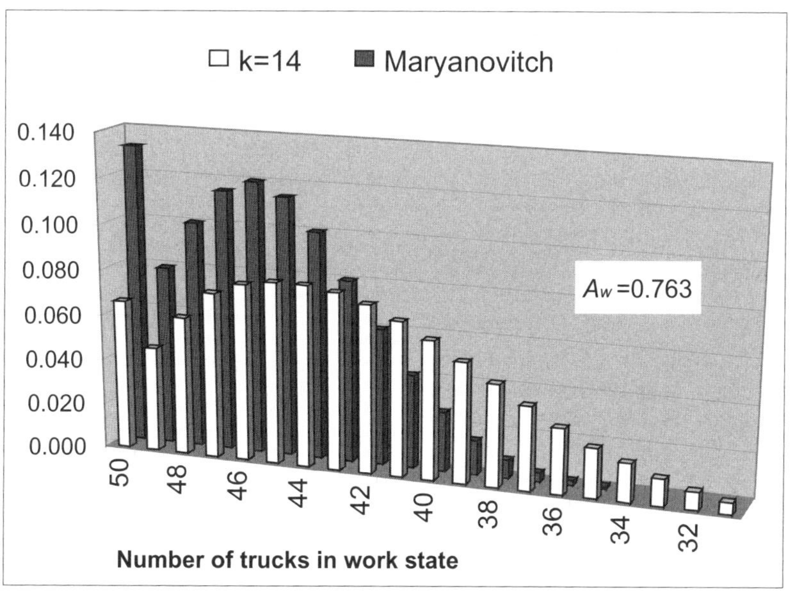

Figure 7.29. The probability distributions of a number of trucks in work state for systems: $\mathfrak{S}_{IX}$ and $\mathfrak{S}_{X}$.

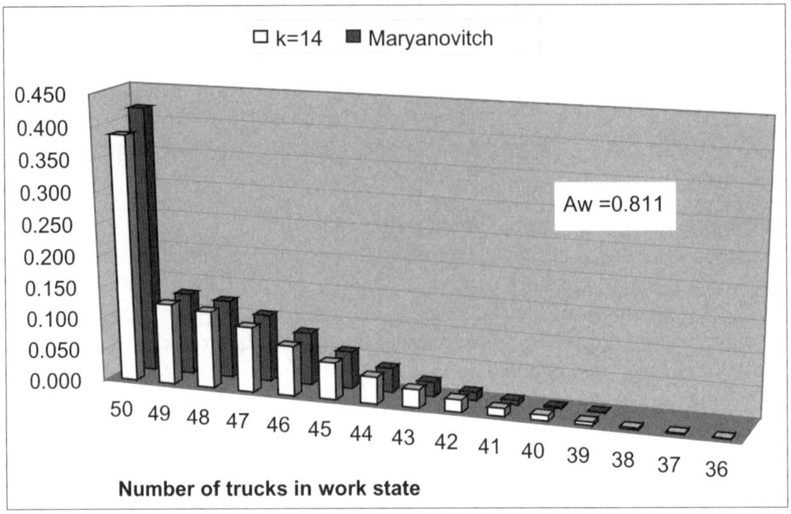

Figure 7.30. Probability distributions of numbers of trucks in work state for systems: $\mathfrak{S}_{IX}$ and $\mathfrak{S}_{X}$ with high steady-state availability of trucks.

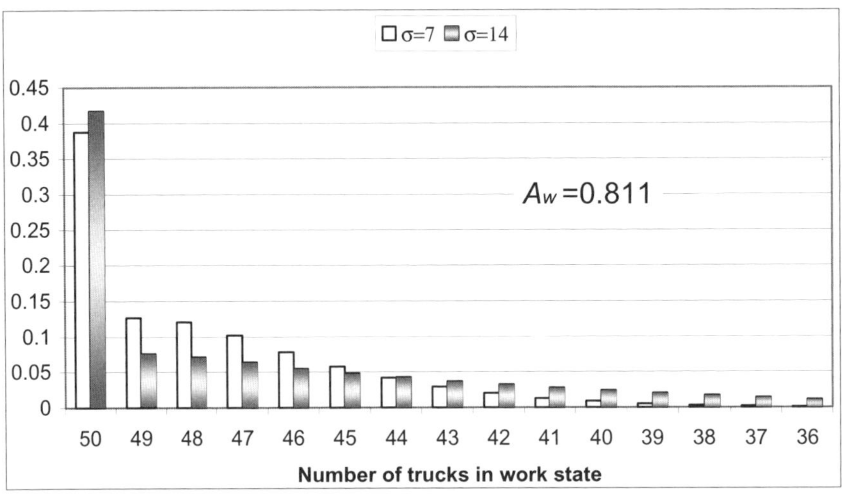

Figure 7.31. Probability distributions of numbers of trucks in work state for systems $\mathfrak{S}_{IX}$ with trucks of high availability and different standard deviation of repair time σ_n.

Compared to Figure 7.29, changes in the allocation of the masses of probability are predictable. What is important, though, is that the difference between the probability masses in Figure 7.30 is small. Firstly, it is necessary to state that a decrease in changes of the masses of probability is caused by an increase in the availability of machines. For systems with trucks of high reliability, a smaller number of repair stands is enough to obtain a system with service saturation.

To make the consideration more comprehensive, changes in the distribution of the mass of probability due to changes in standard deviation need to be discussed. Figure 7.31 presents these changes for different values of the standard deviation of times of repair: $\sigma_n = 7$ h and $\sigma_n = 14$ h (meaning that in the second case there is a significantly worse quality of repairs) for trucks of $A_w = 0.811$ (the mean time of repair was assumed to be 7.8 h). It is clear that increasing the standard deviation makes a displacement of the mass probability towards the inception of the coordinate axes, making the whole distribution wider, apart from the point of full output of the truck system.

It is essential to keep in mind the fact that the probability distribution of the number of trucks in work state (7.27) is the input information in a further stage of modelling, and this distribution comprises information on:

- Reliability of trucks
- Organization of the truck system: division between trucks directed to accomplish the transportation task and trucks in reserve
- Size of the repair shop, giving information on possible truck queues at the shop.

For this reason, note that this distribution contains very rich information on the machinery system.

It is important here to notice that now it is possible to verify the *goodness* of selection of the basic system parameters $< m, r, k >$, i.e. the number of trucks directed to work m, the number of trucks directed to reserve r and the number of repair stands k. Recall: these three parameters are structural system parameters.

This problem is now discussed comprehensively.

CHAPTER 8

Verification of selection of structural parameters of the system

At the very beginning it should be stated that the principle of selection of the structural system parameters for a mine at the design stage will not be discussed. The scope of the reasoning here will be the following situation.

A mine is considered, which has a certain machinery system employed to realize mineral production, i.e. the machinery system is known and the exploitation of the mineral runs for a certain period: a period long enough to estimate basic system parameters. If so, it is possible to assess the *goodness* of the selection of the machinery system applied and to make the necessary changes if needed. The *goodness* of a new selection can be estimated relatively quickly.

Generally, the selection of the machinery system under consideration is a point in the following procedure.

Knowing the geological and mining conditions of a mine being designed and creating the mine planning, i.e. planning the sequence of mine development, the selection of a rock extraction system is made. For open pits, blasting according to the properties of the material excavated will do the extraction. Hence, a system of drilling machines has to be chosen to accomplish the blasting task. This system, with a set sequence of work, will generate a sequence of loading points in which portions of broken rock masses wait for removal. This means that a system of loading machines is needed to load the extracted material on to hauling machines. Therefore, a system of shovels is selected whose output is the hauling task for the system of transporting machines. Knowing the parameters of the chosen shovels, a truck is selected to cooperate with the loading units and then the appropriate number of these hauling machines. The capacity of the chosen truck should be in a certain proportion to the capacity of the bucket/dipper of the shovel. Until the 1980s, the minimum number of buckets to load a truck was usually fixed at four. Later some mines decided to apply shovels that loaded a dump truck by three buckets. Today, calculations are based on the application of loading machines able to load a truck in two passes in some cases. It is assumed that both the type of truck and its payload have been selected. The problem now arises of how to decide the number of hauling units with the division into the number of dumpers that will transport the mineral extracted and the rest of the truck fleet needed to haul the excavated waste material.

The number of trucks needed to accomplish the transportation task is usually determined (Caterpillar 1996, Terex Manual 1981, Surface Mining 1990, Czaplicki 2004) from the formula:

$$h = z_t / u_t \tag{8.1}$$

where:

h—required number of trucks

z_t—given transportation task in unit of time

u_t—hauling capacity of selected truck in unit of time.

Denoting as W_{mxef} the maximum effective output[1] of the selected shovel, as Q the truck payload and as T_c the truck cycle time, the following formula can be recorded:

[1] The maximum effective output of the shovel is the output given by formula: $W_{mxef} = Q\, B_k\, A_k / Z'$. This output can be achieved if there is constantly a certain truck to be loaded.

$$h = \frac{nW_{mxef}}{Q/T_c} = n\frac{Q}{Z'}B_k A_k \frac{T_c}{Q} = nG_k \frac{Z' + O + W + R}{Z'} = nG_k\left(1 + \varpi^{-1}\right) \tag{8.2}$$

where:

B_k—the accessibility coefficient of the shovel
A_k—the steady-state availability of the shovel
G_k—the shovel loading capacity coefficient
n —the number of loading shovels
Z' —the adjusted mean time of loading
O —the mean time of truck haulage
W—the mean dump time of the truck
R —the mean time of truck return from dumping point to shovel
ϖ —the coefficient of relative intensity of loading given by the formula:

$$\varpi = \frac{Z'}{O + W + R}. \tag{8.3}$$

The required number of trucks depends on four parameters of the system B_k, A_k, n and Z', and on three parameters connected with trucks O, W and R. It can be assumed that the fourth truck parameter here is the mean time Z'. This parameter is common to both systems.

Notice immediately that if a system has only one shovel and if this machine is continuously accessible and totally reliable, then the number of trucks required determines the number describing how many times the truck mean time work cycle is greater than the mean time of loading.

Analyzing formula (8.2) more carefully, the following remarks can be made:

- The required number of trucks does not depend directly on the truck payload Q that looks incorrect
- Properties of the loading shovels are represented by two parameters—B_k and A_k—but these parameters are lacking in relation to hauling trucks
- The formula of the mean time of the truck work cycle does not contain any delays, i.e. no truck queues at the shop, no queues at the shovel, etc
- The two parameters O and R depend on the stage of mine development, or to put it more precisely, on the hauling distance L; that means the number of trucks required is a function of L.

The above remarks are discussed in sequence.

The magnitude Q, due to mathematical simplification, has vanished, but it is still there. At first, it was taken into account when the problem of the correct selection of the bucket capacity in relation to the truck payload was considered. This choice had an influence on the parameter Z', a component of formula (8.2).

Czaplicki (1997) stated that 'the required number of trucks determined by formula (8.2) means in fact the expected value of number of trucks in work state'. In other words, the number h determines h ideal (totally reliable) machines. If their reliability is taken into consideration the number of trucks needed is given by the formula:

$$V = \frac{h}{A_w} = \frac{E(\boldsymbol{D})}{A_w} \tag{8.4}$$

(cf. formula (11) p. 727, Surface Mining 1990)
where: $E(\boldsymbol{D})$—the expected value of the number of trucks in work state
A_w—the steady-state availability of the truck.

At this point a certain doubt may arise. There are two steady-state availabilities, A_w and A'_w. Which one should be used?

Note that for an estimation of parameter h determined by formula (8.2) no information on the pair $< m, r >$ is needed.

By using the unadjusted parameter A_w, no information on the pair is introduced. Moreover, the number V does not take into account a possible truck queue at the shop. But a machinery system should be designed where a possible queue of trucks waiting for repair should happen rather seldom.

By using the adjusted steady-state availability, information on the pair $< m, r >$ is introduced (parameter ϕ is the function of this pair, see formula (7.16)). Therefore, the number of trucks required as the function of the applied number of trucks is obtained. This is a logical error.

The following considers the problem of truck accessibility. During the production shift a truck is out of operation—unable to transport—for a certain time due to many reasons. These may be fuelling, change of drivers, coffee breaks, etc. It has been assessed (e.g. Church 1981, Caterpillar Handbook of Earthmoving 1981, Kolonya *et al.* 2003) that assuming, say, an 8-hour shift, for approximately 1 hour to 1.33 hours the truck will be excluded from operation. This means that there is a state in the process of truck exploitation that absorbs machines and in which they are unable to haul. This state does not absorb trucks in reserve and does not absorb failed trucks. It is therefore necessary to increase the number of trucks in order to fill the gap. This 'lost' number of trucks is 0.125 to 0.167 of the number of trucks directed to work—m. Hence, the recommended number of transporting units will be, say, 1.15 m.[2] These additional trucks can be termed *surplus trucks*. This number of trucks should be in reserve and should be directed to work replacing trucks that are actually going out of operation, but not due to failure.

From a mathematical point of view, the effect of these surplus trucks is as if the truck mean time work cycle increases:

$$T'_c = 1.15T_c.$$

Hence, taking into account trucks' accessibility the following repercussions can be noted:

- Increase in the number of trucks needed for the system
- Increase in the repair stands needed
- Effectively an increase in the truck mean time work cycle
- Increase in the flow rate in the system; the heavy traffic condition is easier to fulfil.

A truck may be in a queue of units waiting for repair or in a queue of trucks waiting to be loaded at the shovel. For the given system structural parameters, and knowing the intensities δ, γ and the standard deviations σ_p and σ_n, the mean time a truck spends waiting for repair can be assessed. This can then be lumped with the state of repair. Thus, the possible occurrence of a queue of trucks waiting for repair is taken into account when the agglutination of states is being made. Information about this fact is included in the adjusted steady-state availability and in the probability distribution of the number of truck in work state. However, at this point the queue of trucks waiting for repair is not taken into account. Note that it is advisable to have a queue of trucks waiting for loading (the best is the shortest possible—one truck) because a shovel should work round-the-clock, especially such huge units. A queue of trucks waiting for repair is in every respect disadvantageous. The problem remains of how to take into account a queue of trucks waiting for loading. Here two solutions are possible. One solution relies on the increase of the mean time truck work cycle, say on a certain small value Δ. The second solution relies on the assumption that the calculated number of trucks needed for the system is the minimum value. Therefore, this rate is increased by a certain amount, expressed usually by a percentage of the computed value. Avoiding here a long discussion on the base for a given amount (which is given by Czaplicki 2006 p. 74) it can be assumed that it is sufficient for the number of trucks needed to be increased by 10%. This is a typical engineering approach, but three reasons also provide a basis in the statistical field. First, looking at formula (8.2) it is easy to notice that one parameter is the determined value n; the rest have to be estimated. In practice, their assessment will be done based on a certain sample. Thus, the estimations obtained will

[2] If in a particular mine a slightly different assessment of this absorbing state (inaccessibility) of trucks is made, e.g. 1.12 or 1.18, this number should be used further on in the modelling.

have a certain correspondence with the real values only. Second, the fact that shovels move from time to time from one loading point to another, and the mean hauling time and return time change should be considered. In order to avoid continuous modifications of the appropriate means, approximate values are assumed and applied in calculations and analyzes. Third, the method of calculation applied here is based on a diffusion approximation and thus—as has been stated—all parameters and functions obtained using this mathematical tool are approximate magnitudes only.

If in later stages of the modelling and analysis the precise number h (8.2) is applied, the final figures obtained would be *too accurate, too sparing*. Therefore, henceforth the number of trucks needed, as calculated by means of formula (8.2), will be enlarged by 10%.

One final remark remains to consider, concerning the changing length of transportation. This problem will be discussed comprehensively in chapter 15. For consideration here, a constant value of O and R is assumed.

To summarize:

The machinery system is given—i.e. the number of shovels applied n, the number of trucks directed to work m and the number of trucks in reserve r are known. The system is in operation and an estimation of basic functioning parameters has been done—parameters Z', O, W *and* R are known. Additionally there are estimates of the steady-state availabilities of machines and their accessibility is known. Therefore, the number of trucks required for the system can be found from the formula:

$$\nabla(V) = \nabla\left(G_k n \frac{T_c}{Z'} \frac{1}{A_w} \right), \nabla(V) \in \mathfrak{R}, \tag{8.5}$$

where symbol ∇ means the necessity to round up the value obtained to the nearest natural number. The natural number defined by formula (8.5) means the number of trucks of determined reliability that should be directed to accomplish the transportation task given.

Knowing the required number of trucks, means that the number of spare trucks can be determined.

Analyzing the literature from this field, two approaches can be identified. The SME Mining Engineering Handbook (1973), Terex Manual (1981), Polish Mining Engineering Handbook (1982) and Hartman (1987) present the principle that spare units should replace failed trucks during their operation. Therefore, the aim of the reserve is to fill up the truck shortage. The principle appears rational but has not been thought through thoroughly. An obvious question arises here: to what extent should this shortage be filled up? The number of failed machines is a random variable. Sometimes there is no failed machine. Sometimes, too, all machines could be down, although the probability of this occurring is usually very small; it decreases with an increase of the number of applied machines as well as with an increase of the reliability of the equipment used. Proofs of the lack of comprehensive analysis in this regard are the methods of calculation of reserve size: different methods are presented in different publications. Moreover, there is no relation at all between the reserve size and the formulated goal of application of transporting machines. There is also—as far as the author knows—no relation at all between the reserve size and the number of repair stands employed.

Czaplicki presented a different approach to the problem of reserve size determination in several publications that were summed up in his 2004 textbook. The departure point in this field is the assumption that the spare units are wanted for two reasons—maintenance needed for operating machines and failures clearing needed in these machines—but all these problems should be considered while keeping in mind the continuous accomplishment of the transportation task. It was observed that the number of trucks calculated from formula (8.5) is over-dimensioned because it means the number of trucks directed to work. The key point is that some units can be withdrawn to reserve and there will still be the expected value of the number of trucks in work state not less than needed. Therefore, it is necessary to investigate the maximum number of trucks that can be withdrawn, while continuing to fulfill the condition that the mean number of trucks in work state is not less than that required. In this way, the problem of the reserve is equivalent to the problem of searching for the appropriate pair $< m, r >$. It has also been noted that there is a certain number of such pairs that accomplish the transportation task given. At this moment, the problem arises over the formulation of a criterion that will allow one to be chosen from the set of all pairs given. At first glance, the economic principle appears to be the most appropriate. Czaplicki (1994) worked on the assumption that the best is the pair for which a profit obtained from its application attains the maximum. The result of these deliberations was unfavourable—there is no way to estimate parameters of the criterion function in mine practice. Therefore, a different, two-stage criterion was used:

a. The total number of trucks applied in the system to accomplish the transportation task given should be minimum
b. From all pairs fulfilling condition (a) the pair that has the maximum reserve must be looked for.

The above means that the following system is being sought:

$$\mathfrak{S} : < \mathfrak{T}, A_w, m, r; (m + r)_{min}, r_{max} \mid E(\boldsymbol{D}) \geq h >, \tag{8.6}$$

where symbol $\mathfrak{T}$ denotes the truck type applied.

The rationale of condition (b) is supported by the following regularities:

- Increment of the reserve has an advantageous influence on the survivability of machines because they will be in reserve for longer after repair; lower intensity of their usage, better planning of diagnostics, higher quality of maintenance. The last two are because of lack of time pressure.
- More machines in reserve, less machines in shop, lower probability of queue of machines waiting for repair, lower usage of truck roads, higher safety, lower probability of truck accident.
- Very high unit cost of running a truck; low unit cost of a truck in reserve.
- More trucks in pit, higher probability of their mutual disturbance, more emission gasses in pit, the mean truck work cycle usually increases.

Two remarks to conclude the considerations on the problem of reserve size:

The division of a truck fleet into a pair $< m, r >$ should not be treated as the *constant oracle*. The main point is that the sum $m + r \geq h$. The division makes sense at the moment of regeneration of the exploitation process. The term *regeneration* will be explained in chapter 10. In a relatively short time after regeneration some machines will fail and states of shovel inaccessibility will occur. These reasons provide the pair: the number of machines directed to work and the number of machines in reserve is a two-dimensional random variable where values can change at every moment of time.

Concerning the general approach to the problem of reserve, it should be remembered that the reserve size must fulfil the requirements connected with reliability of circulating machines and with planned and unplanned maintenance of machines. Both components must be taken into account. This problem, though, is beyond the scope of this book.

Having selected the structural parameters of the system $< m, r, k >$ the *goodness* of this selection can be verified. The procedure of verification is shown in Figure 8.1.

The following are three examples of this procedure.

Example 1

Take a system of $n = 12$ shovels. The steady-state availability of a shovel is $A_k = 0.860$ and its accessibility coefficient is $B_k = 0.850$. The steady-state availability of a truck is $A_w = 0.811$. The truck intensity of failures δ has been assessed as $0.030\ h^{-1}$, with the standard deviation $\sigma_p = 33$ h and the intensity of repair $\gamma = 0.129\ h^{-1}$ with the standard deviation $\sigma_n = 6.5$ h.

Estimates of the mean times of truck work cycle phases are as follows: $Z' = 1.9$ min, $O = 15.4$ min, $W = 1.0$ min, $R = 10.3$ min.

Find the structural system parameters for this system.

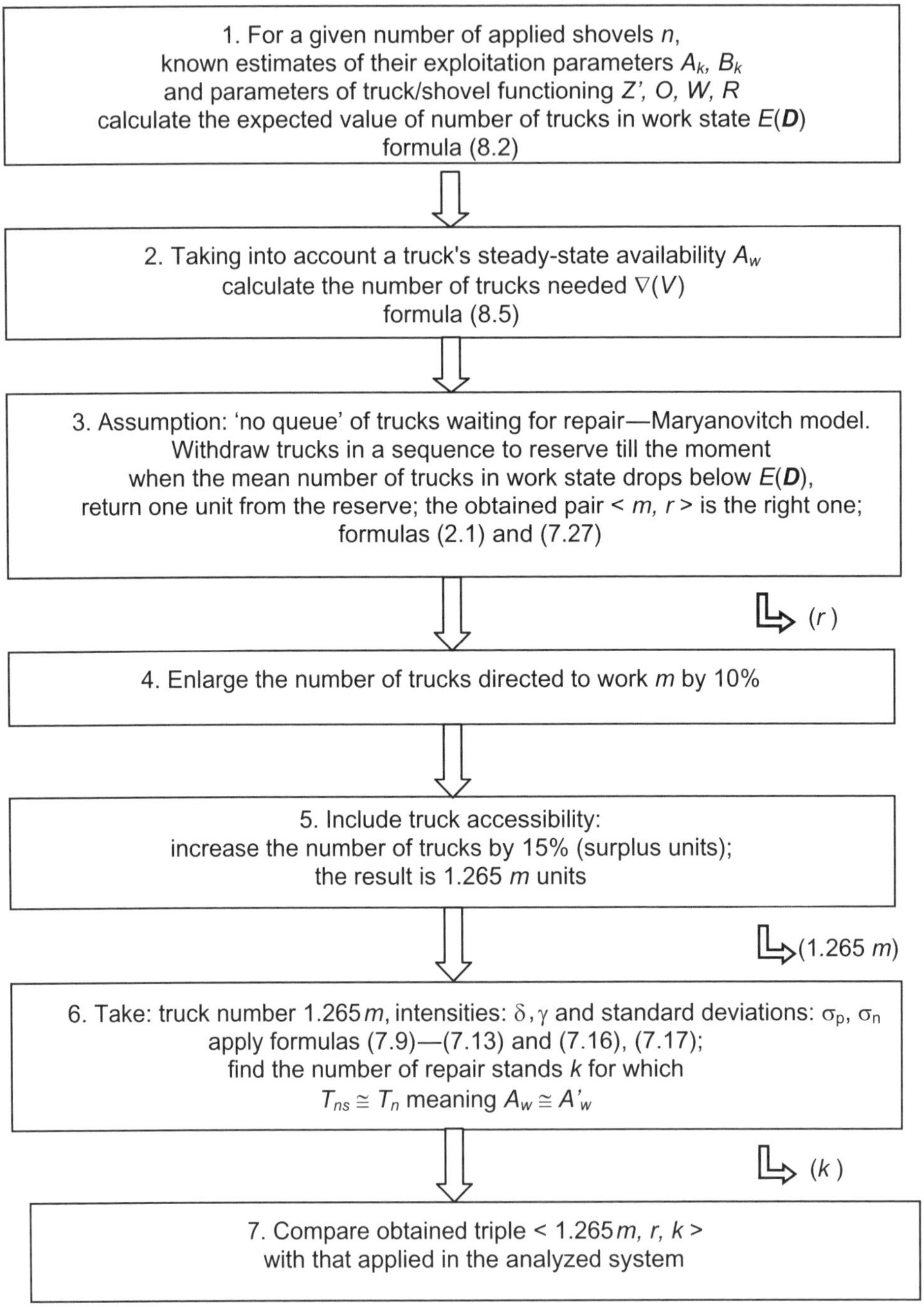

Figure 8.1 Scheme of determination of the system structural parameters.

1. The expected value of the number of trucks in work state is

$$E(\boldsymbol{D}) = nA_k B_k \frac{Z' + O + W + R}{Z'} = 132.04 \text{ trucks}$$

2. The number of trucks needed
   $V = E(D)/A_w = 162.81 \Rightarrow 163$ trucks
3. Apply the Maryanovitch model (2.1).
   Start from the system < 163, 0 >. For this system $E(\boldsymbol{D}) = 132.19$.
   For < 162, 1 > $E(\boldsymbol{D}) = 132.19$, … for < 140, 23 > $E(\boldsymbol{D}) = 132.10$.
   If 1 more truck is withdrawn the mean $E(\boldsymbol{D})$ will be just 132.04.
   So the right system is: < 140, 23 >.[3]
4. Enlarge the number of trucks by 10%, yielding 154 units.
5. Taking into consideration the truck's accessibility add surplus units—15% of the total number. The result is 177 trucks.
6. Applying the set of formulas (7.9)–(7.13) and having formulas (7.16), (7.17) as the measures of matching, it is found that for the number of repair stands $k = 48$, formulas (7.16) and (7.17) hold.
7. The machinery system should characterize the following parameters: < 154 + 23, 24, 48 >, i.e. 154 trucks should be directed to work and 23 additional trucks should replace units going for fuelling, coffee breaks, etc., and a supplementary 24 trucks should create a reserve and the repair shop should possess 48 stands. The truck queue at the shop should be negligible.

Example 2

Take a system of $n = 8$ shovels. The steady-state availability of a shovel is $A_k = 0.860$ and their accessibility coefficient is $B_k = 0.850$. The steady-state availability of a truck is $A_w = 0.781$. The truck intensity of failures δ has been assessed as 0.033 h^{-1}, with a standard deviation $\sigma_p = 30$ h and the intensity of repair $\gamma = 0.118$ h^{-1} with a standard deviation $\sigma_n = 7$ h.

The parameters of the truck's functioning remain intact.

Find the structural system parameters for this system.

1. The expected value of number of trucks in work state is:

$$E(\boldsymbol{D}) = nA_k B_k \frac{Z' + O + W + R}{Z'} = 88.03 \text{ trucks}$$

2. The number of trucks needed is:
   $V = E(D)/A_w = 112.7 \Rightarrow 113$ trucks
3. Apply the Maryanovitch model (2.1).
   Start from the system < 113, 0 >. For this system $E(\boldsymbol{D}) = 88.25$.
   For < 112, 1 > $E(\boldsymbol{D}) = 88.25$, … for < 94, 19 > $E(\boldsymbol{D}) = 88.10$.
   If 1 more truck is withdrawn the mean $E(\boldsymbol{D})$ will be below 88.03.
   The right system is: < 94, $r = 19$ >.
4. Enlarge the number of trucks that will be directed to work by 10%, yielding 104 units.
5. Taking into consideration the truck's accessibility add surplus units—15% of the total number. The result is 120 trucks.
6. Applying the set of formulas (7.9)–(7.13) and having the formulas (7.16), (7.17) as the measures of matching it is found that for the number of repair stands $k = 39$, equations (7.16) and (7.17) hold.

[3] If $m = 140$ trucks are directed to work and there is no truck reserve, this system is equivalent to $mA_w = 113.5$ totally reliable trucks. If $m + r = 163$ trucks are directed to work then taking into consideration their reliability this system is equivalent to 132.2 totally reliable trucks. Withdrawing 23 machines to reserve, the organization of the system is changed, and it is then equivalent to 132.1 totally reliable trucks.

7. The machinery system should characterize the following parameters: $< 104 + 16, 19, 39 >$, i.e. 104 trucks should be directed to work and 16 additional trucks should replace units going for fuelling, coffee breaks, etc., and a supplementary 19 trucks should create a reserve and the repair shop should have 39 stands. The truck queue at the shop should be negligible.

Example 3

Take a system of $n = 4$ shovels. The steady-state availability of a shovel is $A_k = 0.860$ and their accessibility coefficient is $B_k = 0.850$. The steady-state availability of a truck is $A_w = 0.725$. The truck intensity of failures δ has been assessed as 0.041 h^{-1}, with a standard deviation $\sigma_p = 24$ h and the intensity of repair $\gamma = 0.108$ h^{-1} with a standard deviation $\sigma_n = 7.5$ h.

The parameters of the truck's functioning remain intact.

Find the structural system parameters for this system.

1. The expected value of number of trucks in work state is

$$E(\boldsymbol{D}) = nA_k B_k \frac{Z' + O + W + R}{Z'} = 44.01 \text{ trucks}$$

2. The number of trucks needed is:
   $V = E(D)/A_w = 60.7 \Rightarrow 61$ trucks
3. Apply the Maryanovitch model (2.1).
   Start from the system $< 61, 0 >$. For this system $E(\boldsymbol{D}) = 44.23$.
   For $< 48, 13 >$ $E(\boldsymbol{D}) = 44.04$.
   If 1 more truck is withdrawn the mean $E(\boldsymbol{D})$ will be below 44.01.
   The appropriate system is: $< 48, r = 13 >$.
4. Enlarge the number of trucks directed to work by 10%, yielding 53 units.
5. Taking into consideration truck accessibility add surplus units—15% of the total number. The result is 61 trucks.
6. Applying the set of formulas (7.9)–(7.13) and having formulas (7.16), (7.17) as the measures of matching, it is found that for the number of repair stands $k = 28$, equations (7.16) and (7.17) hold.
7. The machinery system should characterize the following parameters: $< 53 + 8, 13, 28 >$, i.e. 53 trucks should be directed to work and 8 additional trucks should replace units going for fuelling, coffee breaks, etc., and a supplementary 13 trucks should create the reserve and the repair shop should have 28 stands. The truck queue at the shop should be negligible.

If these machinery systems are now compared. A summary of their parameters is given below. One system was added in order to put the last two side-by-side; their difference relies on different values of the steady-state availability only.

$\mathfrak{S}_1 : < n = 12;\ A_w = 0.811;\quad m = 154 + 23,\ r = 24,\ k = 48 >$
$\mathfrak{S}_2 : < n = 8;\ A_w = 0.781;\quad m = 104 + 16,\ r = 19,\ k = 39 >$
$\mathfrak{S}_3 : < n = 4;\ A_w = 0.725;\quad m = 53 + 8,\ r = 13,\ k = 28 >$
$\mathfrak{S}_4 : < n = 4;\ A_w = 0.811;\quad m = 51 + 8,\ r = 9,\ k = 26 >$

The Reader may construct his or her own conclusions. The figures are significant. There is no doubt that the reliability of machines has a great influence on the machinery system needed and, later, on the profitability of the whole enterprise. As can be seen, this conclusion is still valid.

Note that when comparing systems 3 and 4, if it is decided to purchase trucks of approximately 10% lower availability 10% (six) additional trucks will be needed as well as 8% (two) more repair stands.

CHAPTER 9

Modelling and analysis of the exploitation process of a shovel-truck system: Part II

9.1 RELIABILITY OF REPAIR STANDS

This problem does not exist in considerations concerning the exploitation characteristics of shovel-truck systems in mining. However, in the theory of reliability in the 1980s many publications looked at the reliability of diagnostic systems, controlling devices and repair equipment.

Arranging a back-up facility for the machinery system considered is very often a great enterprise. It comprises measuring equipment, various instrumentation, tools and manipulators, maintenance and repair procedures, diagnostic systems, spare parts, garbage disposal—especially the problem of the huge quantity of worn tyres, construction of fuel tanks, arrangement of truck bays, etc. This type of arrangement is therefore a great technical and economic project.

The scope of this consideration is confined to random phenomena occurring during the exploitation process of the machinery system as well as the stochastic properties of the equipment involved in restoring the machine's ability to work. It has already been seen how to include the number of repair stands applied in the system in an analysis of the whole system. But one vital problem has been omitted. Repair stands can sometimes be inaccessible. Some such systems operate in a cyclical way, a certain number of hours per day. This is planned, and therefore this case does not merit discussion here. Sometimes stands are out of work because the equipment that comprises part of the stand failed. In such cases, the stands can be inaccessible in a random way. This problem lies within the scope of these considerations.

It is a fact that stands in the shop can fail, i.e. they are characterized by reliability. For the simplest case, where the shop consists of k stands of the same reliability, and if a truck has failed it is directed to any stand. The stands are characterized by a steady-state availability A_n. The system described in this way is a k-element system, its units operate independently of each other, i.e. operate in parallel in the reliability sense. For this system, the probability distribution of the number of stands in work state is given by:

$$P_{ni}^{(p)} = \binom{k}{i} A_n^i (1 - A_n)^{k-i} \qquad i = 0,\ 1,\ 2, \ldots,\ k. \tag{9.1}$$

The expected value of the number of stands in work state is determined by the formula:

$$E_n = k A_n. \tag{9.2}$$

The reliability of such a system is a trivial problem, but much more significant are the repercussions that result from the random access of stands. The Sivazlian and Wang model *G/G/k/r* must be modified to *G/G/**i**/r*, where ***i*** is now the random variable of the known probability distribution (9.1). This means that a determined randomization of the model must be done.

Applying the heavy traffic condition, the probability distribution of the number of failed objects is obtained, and depending on whether the number of repair stands is greater than the number of spare units or not a different set of formulas must be taken into account. Moreover, if the number of repair stands k is not greater than the reserve size r ($k \leq r$) then the set of formulas (2.4) should be applied. However, if the number of repair stands k is greater than the number of spare units r ($k > r$) then both sets of formulas (2.4) and (2.5) have to be taken into account. Some events will occur in which the number of repair stands k will be less than r, equal to r and more than r. This second case is more general than the first one, and will thus be the point of this investigation. Remember that, as previously stated, the number of repair stands is usually greater than the reserve size, $k > r$.

The following is a discussion of this second, more general case.

Using the following notations: i, $i = 0, 1, \ldots, r, \ldots, k$ the number of accessible stands of the system one can find that the probabilities of events where $\boldsymbol{i} \leq r$ are closed by the sum

$$\sum_{i=0}^{r} P_{ni}^{(p)}.$$

For the above case, the probability density function of the number of failed trucks expressed in continuous form is given by (compare formulas (2.4)):

for $0 \leq x < i$:

$$h_1(x) = K_{11}\, g_1(x) = \frac{K_{11}}{m\xi C_M + xC_R}\left(\frac{m\xi C_M + xC_R}{m\xi C_M}\right)^{\beta_1} \exp\left(\frac{-2x}{C_R}\right) \tag{9.3a}$$

for $i \leq x < r$:

$$h_2(x,i) = K_{22}\, g_2(x) = \frac{K_{22}}{m\xi C_M + iC_R} \exp\left(\frac{2(m\xi - i)(x-i)}{m\xi C_M + iC_R}\right) \tag{9.3b}$$

for $r \leq x \leq m + r$:

$$h_3(x,i) = K_{33}\, g_3(x) = \frac{K_{33}}{(m+r-x)\xi C_M + iC_R}\left(\frac{(m+r-x)\xi C_M + iC_R}{m\xi C_M + iC_R}\right)^{i\beta'_3} \exp\left(2(x-r)/C_M\right) \tag{9.3c}$$

where β_1 is determined in formulas (7.9), whereas:

$$\beta'_3 = \frac{2\left[1 + (C_R/C_M)\right]}{\xi C_M}. \tag{9.3d}$$

The following give the conditions of function continuality, defining parameters being functions of the variable $\boldsymbol{i}$:

$$\alpha_1(i) = \frac{h_3(r,i)}{h_2(r,i)} \quad \alpha_2(i) = \frac{h_2(i,i)}{h_1(i)}. \tag{9.4}$$[1]

The following defines the constants K that are now the functions of variable i and probabilities $P_{ni}^{(p)}$.

$$K_{33}(i) = \frac{P_{ni}^{(p)}}{\alpha_1(i)\alpha_2(i)\int_0^i g_1(x)dx + \alpha_1(i)\int_i^r g_2(x,i)dx + \int_r^{m+r} g_3(x,i)dx}$$

$$K_{22}(i) = \alpha_1(i)K_{33}(i) \quad K_{11}(i) = \alpha_1(i)\alpha_2(i)K_{33}(i). \tag{9.5}$$

[1] $\alpha_1(i) \equiv \alpha_{1i}$, however for greater communicativeness, the notation is assumed as in (9.4) and this form will be further employed in relation to the normalization constants K.

All the components are now determined. The probability distribution of the number of failed machines in the system can now be constructed: unreliable trucks—unreliable repair stands.

Given below are only those probabilities that fulfil the inequality $\boldsymbol{i} \leq r$:

$$P_0^{(\leq r)} = \sum_{i=1}^{r} K_{11}(i) \int_{0}^{0.5} g_1(x)dx$$
$$P_1^{(\leq r)} = \int_{0.5}^{1} K_{11}(1)g_1(x)dx + \int_{1}^{1.5} K_{22}(1)g_2(x,1)dx + \sum_{i=2}^{r} K_{11}(i) \int_{0.5}^{1.5} g_1(x)dx$$
$$P_2^{(\leq r)} = \int_{1.5}^{2.5} K_{22}(1)g_2(x,1)dx + \int_{1.5}^{2} K_{11}(2)g_1(x)dx + \int_{2}^{2.5} K_{22}(2)g_2(x,2)dx + \sum_{i=3}^{r} K_{11}(i) \int_{1.5}^{2.5} g_1(x)dx \quad (9.6)$$

and so on.

The major barrier to presenting the above probability function in an elegant form is the *moving* point of contact of functions $h_1(x)$ and $h_2(x)$ and further $h_2(x)$ and $h_3(x)$ in relation to the number of accessible repair stands.

If, for example, for a system of $k = 12$ unreliable repair stands, being the back-up for m trucks directed to work and $r = 10$ spare trucks, then the probability distribution of the number of failed trucks is given by:

$$P_3^{(\leq r)} = \int_{2.5}^{3.5} K_{22}(1)g_2(x,1)dx + \int_{2.5}^{3.5} K_{22}(2)g_2(x,2)dx + \int_{2.5}^{3} K_{11}(3)g_1(x)dx + \int_{3}^{3.5} K_{22}(3)g_2(x,3)dx$$
$$+ \sum_{i=4}^{r} K_{11}(i) \int_{2,5}^{3.5} g_1(x)dx$$

$$P_4^{(\leq r)} = \sum_{i=1}^{3} K_{22}(i) \int_{3.5}^{4.5} g_2(x,i)dx + \int_{3.5}^{4} K_{11}(4)g_1(x)dx + \int_{4}^{4.5} K_{22}(x,4)dx + \sum_{i=5}^{r} K_{11}(i) \int_{3.5}^{4.5} g_1(x)dx$$

$$P_5^{(\leq r)} = \sum_{i=1}^{4} K_{22}(i) \int_{4.5}^{5.5} g_2(x,i)dx + \int_{4.5}^{5} K_{11}(5)g_1(x)dx + \int_{5}^{5.5} K_{22}(x,5)dx + \sum_{i=6}^{r} K_{11}(i) \int_{4.5}^{5.5} g_1(x)dx$$

$$P_6^{(\leq r)} = \sum_{i=1}^{5} K_{22}(i) \int_{5.5}^{6.5} g_2(x,i)dx + \int_{5.5}^{6} K_{11}(6)h_1(x)dx + \int_{6}^{6.5} K_{22}(x,6)dx + \sum_{i=7}^{r} K_{11}(i) \int_{5.5}^{6.5} g_1(x)dx$$

$$P_7^{(\leq r)} = \sum_{i=1}^{6} K_{22}(i) \int_{6.5}^{7.5} g_2(x,i)dx + \int_{6.5}^{7} K_{11}(7)g_1(x)dx + \int_{7}^{7.5} K_{22}(x,7)dx + \sum_{i=8}^{r} K_{11}(i) \int_{6.5}^{7.5} g_1(x)dx$$

$$P_8^{(\leq r)} = \sum_{i=1}^{7} K_{22}(i) \int_{7.5}^{8.5} g_2(x,i)dx + \int_{7.5}^{8} K_{11}(8)g_1(x)dx + \int_{8}^{8.5} K_{22}(x,8)dx + \sum_{i=9}^{r} K_{11}(i) \int_{7.5}^{8.5} g_1(x)dx$$

$$P_9^{(\leq r)} = \sum_{i=1}^{8} K_{22}(i) \int_{8.5}^{9.5} g_2(x,i)dx + \int_{8.5}^{9} K_{11}(9)g_1(x)dx + \int_{9}^{9.5} K_{22}(x,9)dx + K_{11}(10) \int_{8.5}^{9.5} g_1(x)dx$$

$$P_{10}^{(\leq r)} = \sum_{i=1}^{9} \left(K_{22}(i) \int_{9.5}^{10,} g_2(x,i)dx + K_{33}(i) \int_{10}^{10.5} g_3(x,i)dx \right) + K_{33}(10) \int_{10}^{10.5} g_3(x,10)dx$$

$$P_j^{(\leq r)} = \sum_{i=1}^{r-1} K_{33}(i) \int_{j-0.5}^{j+0.5} g_3(x,i)dx \quad j = 11, \ldots, m+r-1$$

$$P_{m+r}^{(\leq r)} = \sum_{i=1}^{r} K_{33}(i) \int_{m+r-0,5}^{m+r} g_3(x,i)dx.$$

Recall that the whole mass of probability described by formulas (9.6) is $\sum_{i=0}^{r} P_{ni}^{(p)}$.
The rest of the mass is given (compare formulas (2.5) by:

for $0 \le x < r$

$$K_{44} g_4(x) = K_{11} g_1(x) \tag{9.7a}$$

for $r \le x \le i$

$$K_{55}\, g_5(x) = \frac{K_{55}}{(m+r-x)\xi C_M + xC_R} \left(\frac{(m+r-x)\xi C_M + xC_R}{m\xi C_M + rC_R} \right)^{\beta_5} \exp\left(\frac{2(\xi+1)(x-r)}{\xi C_M - C_R} \right) \tag{9.7b}$$

for $i \le x \le m+r$

$$K_{66}\, g_6(x,i) = \frac{K_{66}}{(m+r-x)\xi C_M + iC_R} \left(\frac{(m+r-x)\xi C_M + iC_R}{(m+r-i)\xi C_M + iC_R} \right)^{i\beta'_3} \exp\left(2(x-i)/C_M \right) \tag{9.7c}$$

whereas $i = r+1, \ldots, k$.

The conditions of function continuity are obtained:

$$\alpha_3(i) = \frac{h_6(i,i)}{h_5(i)} \quad \alpha_4 = \frac{h_5(r)}{h_4(r)}. \tag{9.8}$$

The constants K are determined by:

$$K_{66}(i) = \frac{P_{ni}^{(p)}}{\alpha_3(i)\alpha_4 \int_0^r g_1(x)dx + \alpha_3(i) \int_r^i g_5(x)dx + \int_i^{m+r} g_6(x,i)dx}$$

$$K_{55}(i) = \alpha_3(i) K_{66}(i) \quad K_{44}(i) = \alpha_3(i)\alpha_4 K_{66}(i). \tag{9.9}$$

For $\boldsymbol{i} > r$, the following set of formulas are obtained:

$$\begin{aligned}
P_0^{(>r)} &= \sum_{i=r+1}^{k} K_{44}(i) \int_0^{0.5} g_1(x)dx \\
P_j^{(>r)} &= \sum_{i=r+1}^{k} K_{44}(i) \int_{j-0.5}^{j+0.5} g_1(x)dx \qquad \text{for } j = 1, \ldots, r-1 \\
P_r^{(>r)} &= \sum_{i=r+1}^{k} \left(K_{44}(i) \int_{r-0.5}^{r} g_1(x)dx + K_{55}(i) \int_r^{r+0.5} g_{55}(x)dx \right) \\
P_{r+j}^{(>r)} &= \sum_{i=r+1}^{k} K_{55}(i) \int_{r+j-0.5}^{r+j+0.5} g_{55}(x)dx \qquad \text{for } j = 1, \ldots, k-r-1, \\
P_k^{(>r)} &= \sum_{i=r+1}^{k} (K_{55}(i) \int_{k-0.5}^{k} g_{55}(x)dx + K_{66}(i) \int_k^{k+0.5} g_{66}(x,i)dx) \\
P_{k+j}^{(>r)} &= \sum_{i=r+1}^{k} K_{66}(i) \int_{k+j-0.5}^{k+j+0.5} g_{66}(x,i)dx \quad \text{for } j = 1, \ldots, m+r-1 \\
P_{m+r}^{(>r)} &= \sum_{i=r+1}^{k} K_{66}(i) \int_{m+r-0.5}^{m+r} g_{66}(x,i)dx.
\end{aligned} \tag{9.10}$$

Thus, the probability distribution of the number of failed machines P_j is given by:

$$P_j = P_j^{(\le r)} + P_j^{(>r)} \quad \text{for } j = 0, 1, \ldots, m + r. \tag{9.11}$$

To summarize:

The reliability of the repair stands of the shop are being considered. It was assumed that the system of stands is equivalent to the k-element system of units operating independently of each other, meaning in terms of reliability, units working in parallel. The reliability of stands is described by the steady-state availability A_n. A stream of failed trucks is directed to the shop. The probability distribution of the number of failed machines being repaired by the shop equipped in stands that can fail is given by formula (9.11). To transform this probability distribution into the probability distribution of the number of trucks in work state following relation (7.27) does not cause any problem. Remember that this distribution is the input information in a later stage of modelling and finally during calculation of the system.

Failing stands strengthen the heavy traffic situation, a phenomenon that is disadvantageous for the efficiency of the functioning of the whole system. This phenomenon is advantageous, in turn, for the fulfilment of the heavy traffic condition assuring a good estimation of probability distributions.

Mines with large, expensive machinery systems do all they can to keep these stands in good condition. As a rule, trucks need a lot of maintenance, and any failures that occur consume much time and labour. Therefore, the probability that a stand is unable to repair a truck is relatively low. If this is the case, the problem of unreliable stands can be simplified by assuming a certain correction coefficient that takes into account the disadvantageous property of the shop. Looking at the problem of the assurance of continuous work of the hauling system the following solution can be proposed: increase the number of trucks in operation, knowing that some of these units absorb a queue at the shop. The solution is simple but costly. Moreover, an increase in the number of running trucks will increase the number of failed machines and the size of the queue. The second solution is an increase in the number of repair stands. This solution is sometimes expensive and sometimes hard to achieve. However, it seems that the latter proposition is the more rational one.

In a case when the repair stands are really unreliable, the probability distribution that should be used in further modelling and analysis should be that determined by formula (7.27) for probabilities given by formula (9.11).

However, the *destination* probability distribution indicated by the theoretical and empirical investigations should be the distribution obtained from the Maryanovitch model. This means that the machinery system should be organized in such a way that the average queue of failed trucks waiting for repair at the shop should be negligible. And it is this probability distribution that is assumed in the further considerations.

9.2 SHOVEL-TRUCK SYSTEM

The input information required in further modelling and analysis could be summarized as follows:

- The number of shovels n
- Reliability of machines represented by the steady-state availabilities of loading machines A_k and of hauling trucks A_t
- The probability distribution of number of shovels able to load $P_{kd}^{(zd)}$
- Truck fleet parameters, that is:
  - The number of trucks directed to work m increased by 10%
  - The number of surplus trucks (+15%)
  - The number of trucks in reserve r
- The number of repair stands k
- The probability distribution of the number of trucks in work state $P_{wb}^{(p)}$
- Mean times of truck work cycle phases; Z', O, W, R estimated for the whole system.

A shovel system characterized by the probability distribution $P_{kd}^{(zd)}$ generates the stream of mass that is taken by the truck system characterized by the probability distribution $P_{wb}^{(p)}$. Transporting machines circulate between loading points determined by the actual positions of shovels and unloading points. The latter have their positions either in the peripheral devices of a preparation plant or at dumps where waste is located waiting for the end of mineral exploitation. Hauling machines travel between these points according to the dispatching rules implemented and adjusted by the decisions of the truck dispatcher, who is in constant touch with the operation situation in the pit. With one eye fixed on the situation in the pit, the dispatcher continuously checks the positions of machines and their states. Generally, he tries to avoid situations where a shovel is waiting for a truck and the occurrence of truck queues at shovels.

It is assumed that the mine is arranged in such a way that the probability of the development of a truck queue for unloading is negligible.

The goal of the considerations in this chapter is modelling of the exploitation process of the machinery system, so that the basic and auxiliary system parameters can be found. These will create the basis for a comprehensive assessment of the efficiency of the system[2]. The decisions of the truck dispatcher should also be taken into account because his decisions affect the situation in the pit.

Taking into account the fact that originally the truck work cycle is two-stage, the queue system can be investigated as a four-phase one and every particular phase considered as a certain stage of service—Figure 9.1. The stream of arrivals to a particular stage is the output stream from the previous stage. And the output stream from a stage is the input data for the next. However, considering that the times of service on a particular stage have approximately normal probability distribution, the number of clients changes randomly, and each stage possesses a different number of service stands, also randomly changed in certain limits, thus making a construction of the queue system of such properties a very difficult task. Moreover, a procedure of analysis and calculations would be tough and tedious. The communicativeness of the whole considerations would be weak, with a wide range of necessary simplifications and approximations. The practical usefulness would be very doubtful. Modelling and analysis should be communicative and easy to understand. Therefore, it is necessary to modify the consideration of the exploitation process of the system and avoid further simplifications and approximations.

For that reason, the truck work cycle should be considered as a cycle consisting of two phases: service (loading or unloading) and the rest (haulage–dump–return or return–load–haulage)—Figure 9.2.

Notice that the times of phases can be approximately described by the normal distributions, therefore the sum of phases can be described by the normal distribution of appropriately modified statistical parameters. In this monograph, a queue model that Kendall notation starts with *G/G/* is generally used. Because it has been assumed that there is a no queue at unloading points, the *rest* phase consists of haul–dump–return. This part of the truck route will be referred to as *travel*.[3]

It appears at first glance that the Sivazlian and Wang model is the correct one to describe this two-stage system. This is true, but some modifications must be made, and some parameters must be defined in a different way.

First, the term *repair* is replaced by *service* (here: loading) and the term *work* is replaced by *travel*. Therefore, *a queue of trucks waiting for repair* becomes *a queue of trucks waiting for loading*.

Secondly, repair stands working independently in parallel should be read as shovels working independently in parallel. Thus $k \equiv n$.

---

[2]In many publications concerning technical systems the expression calculation of system can be found. This is a certain mental shortcut meaning that for the basic system its parameters and characteristics have been determined and computed.

[3]Incidentally, there was one system for which the bottleneck was just the unloading point. Several researchers analyzed this system (Barnes *et al.* 1979, Barbaro and Rosenshine 1986, Czaplicki 1997) but it was an exception to the rule.

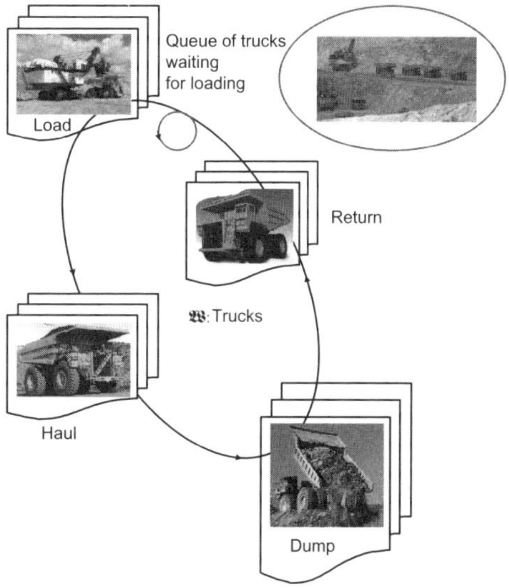

Figure 9.1. Operating scheme of a shovel-truck system—four phases of operation.

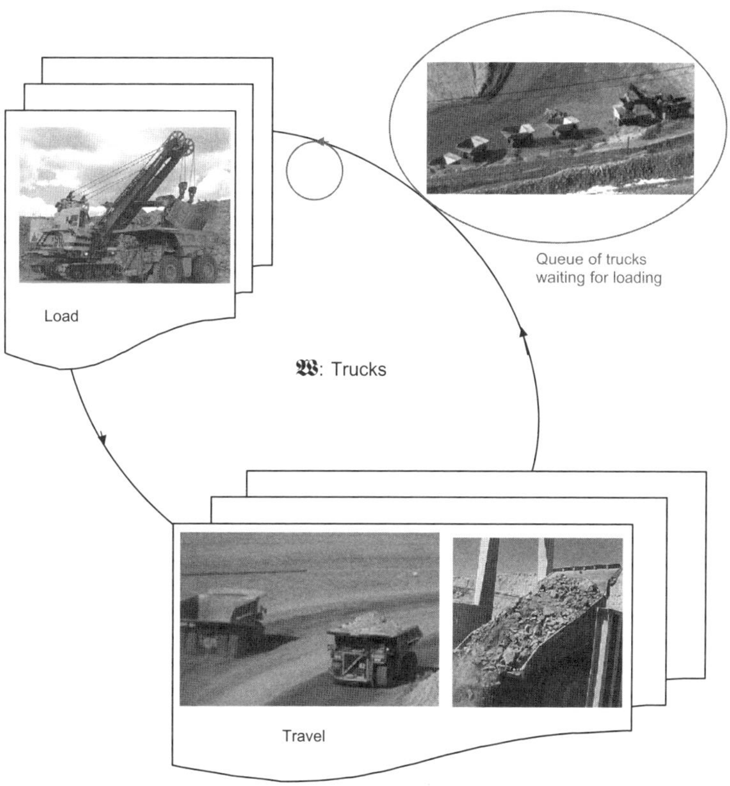

Figure 9.2. Operating scheme of a shovel-truck system—two phases of operation.

For these changes, the model parameters must be defined in a new way. So the following need to be determined:

$$C_1 = \left(\frac{\sigma_j}{T_j}\right)^2 \quad C_2 = \left(\frac{\sigma_z}{Z'}\right)^2 \quad T_j = O + W + R \tag{9.12}$$

where:

σ_j—the standard deviation of times of travel
σ_z—the standard deviation of times of loading.

The last parameter that has to be replaced is the failure rate κ. The new parameter is ϖ and was defined by formula (8.3). This is the coefficient of the relative intensity of loading.

The model in which machines operate continuously, executing their two-phase working cycle now needs to be considered. There is no information in this model on the inaccessibility of trucks or, to put it another way, on the state that absorbs these machines. But it is known that this state exists and must be taken into account. A solution where the time of truck travel is extended would be incorrect. This state is only included by the modification of the number of trucks. The applied number of trucks in the system is 1.1 *m* plus 15% (surplus units). But these surplus units will be lost due to the absorbing state. The result is that these 15% trucks have disappeared, and therefore for further consideration, only 1.1 *m* machines will be assumed.

In the Sivazlian and Wang model the number of machines considered is constant—a deterministic value. But one of the cardinal statements of the theory of machinery systems is that the number of system units in work state is a random variable. Accordingly, it must be assumed that both that the number of shovels in the state of ability for loading is the random variable, and that the number of trucks in work state is the random variable. Thus, there must be a double randomization of the model, but repercussions of this randomization will be different because of the implementation of different dispatching rules and decisions made in connection with the occurrence of failure in a machine of the system. When a truck fails, it is directed to the repair shop, a truck from the reserve is directed to work and the operator changes seats. If there is no truck in the reserve, the driver must wait for a good machine. A different situation occurs when a shovel goes down. The truck dispatcher must make the decision to gradually withdraw a certain number of trucks from the pit to the reserve. If this decision is not made the circulating machines will finish their tour by the shovels in a relatively short time and long truck queues will be observed at each shovel. Trucks will be idly waiting for loading; some transport roads might even be blocked. A better solution is when some trucks are withdrawn from the pit, some vehicles can refuel, some drivers can rest, some trucks can be directed to the shop if needed, and so on. Therefore, it is assumed that for the given number of shovels accessible for loading a different pair number $< m, r >$ will be attributed. This means that as many of these pairs need to ber considered as there are shovels in the system, $< m_d, r_d >$, $d = 1, 2, \ldots, n$. This modification will mean that the parameters obtained during analysis will be conditional ones—provided that *d* shovels are in the state of work. The situation will be analogical to that when randomization was made due to the unreliability of repair stands.

A further modification to the model will be the annulment of the reserve. The reason for this is that information on spare units—on the reserve size and its type—is included in the probability distribution of the number of trucks in work state. It might be assumed that a possible reserve in the model means a surplus of loading points (additional shovels)—but this does not hold. Therefore the considered model will be:

$$G/G/\boldsymbol{d}/m_d$$

where $\boldsymbol{d}$ is the random variable, the number of shovels in work state, whereas m_d denotes the maximum (theoretically) number of trucks waiting for loading in a queue. This variable depends on the number of loading shovels and decisions made by the truck dispatcher. The number m_d does not mean the reserve size as in the classical Kendall notation.

If there is a specified number d, that means that in the machinery system d shovels are in the state of ability for loading. The probability of this event is given by $P_{kd}^{(zd)}$ according to formula (7.4). For this number of shovels there are m_d trucks directed to work and r_d trucks in reserve, according to the decision of the truck dispatcher. One interesting point will be construction of the probability distribution of the number of trucks at shovels. This is the *place* where a queue usually appears. Because $r = 0$, formulas (2.5b), (2.5c), (2.5d) and (7.13), (9.12) need to be taken into account, and determine the probability density function of interesting random variables.

For $0 \leq x \leq d$:

$$K_{55}\, f_5(x;d) = \frac{K_{55}}{(m_d - x)\varpi C_1 + xC_2}\left(\frac{(m_d - x)\varpi C_1 + xC_2}{m_d \varpi C_1}\right)^{m_d \beta'_5} \exp\left(\frac{2(\varpi + 1)x}{\varpi C_1 - C_2}\right) \quad \text{if } \varpi\, C_1 - C_2 \neq 0 \tag{9.13a}$$

$$K_{55}\, f_5(x;d) = \frac{K_{55}}{m_d \varpi C_1}\exp\left(\frac{2x}{C_1} - \frac{(\varpi + 1)\, x^2}{m_d \varpi C_1}\right) \quad \text{if } \varpi\, C_1 - C_2 = 0 \tag{9.13b}$$

For $d \leq x \leq m_d$:

$$K_{66}\, f_6(x;d) = \frac{K_{66}}{(m_d - x)\varpi C_1 + dC_2}\left(\frac{(m_d - x)\varpi C_1 + dC_2}{(m_d - d)\varpi C_1 + dC_2}\right)^{d\beta'_3} \exp\left(\frac{2(x - d)}{C_1}\right) \tag{9.13c}$$

where:

$$\beta'_3 = \frac{2\left[1 + (C_2 / C_1)\right]}{\varpi C_M} \tag{9.14a}$$

$$\beta'_5 = \frac{2\varpi\,(C_1 + C_2)}{(C_2 - \varpi C_1)^2}. \tag{9.14b}$$

These formulas are correct provided that the number of shovels capable of loading continuously equals m_d. But this is not the case. Until now the number of loading shovels randomly changed has been taken into consideration. So, now the number of trucks in work state randomly changed will be included. The parameter m_d has to be replaced by the current number of trucks in work state, $b = 1, 2, \ldots, m_d$; $b \geq x$ and thus the above functions are as follows:

For $0 \leq x \leq d$:

$$K_{55}\, f_5(x;d,b) = \frac{K_{55}}{(b - x)\varpi C_1 + xC_2}\left(\frac{(b - x)\varpi C_1 + xC_2}{b\varpi C_1}\right)^{b\beta'_5} \exp\left(\frac{2(\varpi + 1)x}{\varpi C_1 - C_2}\right) \quad \text{if } \varpi\, C_1 - C_2 \neq 0 \tag{9.15a}$$

$$K_{55}\, f_5(x;d,b) = \frac{K_{55}}{b\varpi C_1}\exp\left(\frac{2x}{C_1} - \frac{(\varpi + 1)\, x^2}{b\varpi C_1}\right) \quad \text{if } \varpi\, C_1 - C_2 = 0 \tag{9.15b}$$

And for $d \le x \le b$:

$$K_{66}\, f_6(x;d,b) = \frac{K_{66}}{(b-x)\varpi C_1 + dC_2}\left(\frac{(b-x)\varpi C_1 + dC_2}{(b-d)\varpi C_1 + dC_2}\right)^{d\beta_3'} \exp\left(\frac{2(x-d)}{C_1}\right). \tag{9.15c}$$

The coefficient taking into account function continuity:

$$\Psi_d(b) = f_6(d;d,b)/f_5(d;d,b), \tag{9.16}$$

Whereas normalization constants are:

$$K_{d66}(b) = \frac{P_{wb}^{(p)}}{\Psi_d(b)\int_0^d f_5(x;d,b)dx + \int_d^b f_6(x;d,b)dx} \qquad K_{d55}(b) = K_{d66}(b)\,\Psi_d(b). \tag{9.17}$$

The probability density function of number x of trucks at d shovels capable of loading given continuous space is:

$$\boxed{f(x;d) = K_{d55} f_5(x;d) + K_{d66} f_6(x;d).} \tag{9.18}$$

As the investigations showed, this probability density function can often be approximated by a normal distribution.

The function determined by formula (9.18) is key to these considerations. It contains rich built-in information on:

- The system of shovels:
  - Number of shovels applied
  - Availability of shovels
  - Accessibility of shovels
  - Reliability structure of shovel system
- The system of trucks:
  - Number of trucks directed to accomplish the transportation task
  - Number of trucks in reserve
  - Type of reserve
  - Reliability of trucks
  - Functioning of trucks, mean times of truck work cycle phases and corresponding standard deviations
  - Information on the system structure depending on the truck dispatcher's decisions
- The system of repair stands:
  - Number of repair stands applied
  - Reliability structure of repair shop
- The decisions made by the truck dispatcher that have an influence on changes in the organization of the truck system.

The last problem to consider is the fulfilment of the heavy traffic condition. It is necessary to check the following inequality:

$$\frac{1.1m}{n}\frac{Z'}{T_j} \ge 0.75. \tag{9.19}$$

Generally, this condition is fulfilled because all the people in the mine know that the shovel should operate round-the-clock, so the arriving trucks keep these loading machines almost always busy. This is the heavy traffic situation.

Commonly, for systems designed properly, i.e. with a relatively large number of transporting trucks to number of shovels, this condition holds. However, for systems with a relatively low number of trucks this inequality could not be fulfilled, therefore application of this procedure can give a poor assessment of the system.

Using the functioning model of the shovel-truck system, the basic and auxiliary system parameters and characteristics can be determined. This allows further analysis of the system and calculation of system parameters such as output of the whole system and its components, its utilization and availability.

CHAPTER 10

Further analysis and system calculation

At this point, almost all of the statistical tools required to construct the essential and auxiliary parameters and characteristics of the system are known. The majority will be measures of system efficiency.

The first step is the construction of the conditional probability of an event that there will not be any truck at d shovels able to load, $d = 1, 2, ..., n$. This probability is determined by the formula:

$$p_{0d} = \sum_{b}^{m_d} P_{wb}^{(p)} K_{d55}(b) \int_{0}^{0.5} f_5(x;d,b)dx. \quad (10.1)$$

Following this, further conditional probabilities are constructed—that there will be g trucks at d shovels able to load:

$$p_{gd} = \sum_{b}^{m_d} P_{wb}^{(p)} K_{d55}(b) \int_{g-0.5}^{g+0.5} f_5(x;d,b)dx \quad \text{for } g = 1, 2, ..., d-1$$

$$p_{dd} = \sum_{b}^{m_d} P_{wb}^{(p)} (K_{d55}(b) \int_{d-0.5}^{d} f_5(x;d,b)dx + K_{d66}(b) \int_{d}^{d+0.5} f_6(x;d,b)dx) \quad g = d$$

$$p_{gd} = \sum_{b}^{m_d} P_{wb}^{(p)} K_{d66}(b) \int_{g-0.5}^{g+0.5} f_6(x;d,b)dx \quad \text{for } g = d+1, ..., m_d \quad (10.2)$$

Observe that:

- Conditional probabilities p_{gd}, $g = 1, 2, ..., d - 1$ are interesting because they provide information about different degrees of imperfect utilization of the shovel system
- Conditional probability p_{gg} is a measure of the most desired event—full matching of both systems: truck and shovel ones
- Conditional probabilities p_{gd}, $g = d + 1, ..., m_d$ are not so interesting; more worthy of note is the sum of these probabilities or—coming back to the continuous space—expression:

$$\sum_{b}^{m_d} P_{wb}^{(p)} K_{d66}(b) \int_{g-0.5}^{m_d} f_6(x;d,b)dx = p_q \quad (10.3)$$

that is the conditional probability of the occurrence of a queue of trucks waiting for load.

Practical hint: if the initial value of these probabilities p_{gd}, $g = d + 1, ...$ is significantly greater than zero it is advisable to check the probability of an event that at one shovel a queue appears that is greater than the allowable.

It is worth taking a moment to enlarge on these considerations while remaining in the practical field.

In the majority of operating systems, the number of trucks is sometimes quite high, but in terms of the number of loading shovels, and bearing in mind too the lengths of transporting routes, this number is usually not very high. The number of trucks in a queue at a shovel therefore often amounts to just a few units, say 2–3, vehicles. The truck dispatcher knows very well the locations of shovels in a pit where the number of allocated trucks should not be too large, and he will not send many trucks there. Consequently, it can be said that there exists a practical limit on the number of g-trucks that can be assigned to a particular place.

Now an important operation parameter must be defined—the conditional expected value of the numbers of trucks at d shovels able to load. Coming from the definition of mathematical hope (5.1), the following is obtained:

$$E_{wkd} = \sum_{b}^{m_d} P_{wb}^{(p)} K_{d66}(b)\left(\Psi_d(b)\int_0^d x f_5(x;d,b)dx + \int_d^b x f_6(x;d,b)dx \right). \tag{10.4}$$

An interesting conditional measure of the shovel service intensity is the ratio of the above expected value to the number of shovels able to load:

$$E_{wkd}/d \tag{10.5}$$

This measure allows an assessment to be made—to a certain degree—whether a pair $< m_d, r_d >$ is properly selected. The study should be made in a penetrating manner and a wider context. If it is assumed that it is good when this ratio is high—significantly above one—then for a large number of working shovels the number of trucks could be quite high and the real truck circulation in the pit can decrease.

Knowing the value of parameter (10.4) an estimate can be made of the conditional mean time lost in the truck work cycle due to waiting in queues for loading. However, before such a measure can be constructed, one subtle problem has to be taken into consideration.

This reasoning was started with the assumption that in the machinery system d shovels are able to load, $d = 1, 2, \ldots, n$. Based on this assumption, several measures of system efficiency, measures of a conditional character have been constructed. In order to escape from this character, i.e. to obtain unconditional measures, it is necessary to multiply the value of a particular parameter by the probability of its appearance and to sum up all such possible cases—meaning to sum up to closed probabilities to unity. Until now, though, one case has been omitted. This is a singular event when all shovels are down or some shovels are down and the rest are in a state of inability to load[1]. The movement of the whole truck system decays in a short time. For this reason in order to obtain the normalization probabilities to unify the appropriate system parameters must be multiplied by the following constant:

$$c = \left(1 - P_{kn}^{(zd)}\right)^{-1}. \tag{10.6}$$

This special case needs some comments.

An event when all shovels, or to put it more precisely, all loading machines in the system, are not able to load is a special event, a singular one during the exploitation process of the system. If the time to finish the closest repair of the shovel (the earliest renewal point in the exploitation process of shovels) or the time to terminate the inaccessibility shovel state is shorter than the truck time work cycle, the whole process of exploitation of the system regenerates itself. In that time all trucks will reach their points of destination, some trucks will be in queues at shovels waiting for loading, and some trucks will be in reserve, some trucks will be in a state of inaccessibility for loading. Truck movement decays. When the process of loading is recommenced and one after another, shovels begin to load, the exploitation process characteristics in this period are different to those when the process is stabilized. It is easy to notice that the fewer shovels in the system, the lower their reliability and accessibility, the higher the probability of occurrence of such a singular state. Czaplicki (2004) described the existence of such a state in the process of the exploitation of a shovel-truck system. From the theoretical point of view, the problem of regeneration of stochastic processes was first noticed by Feller (1949). Kendall (1953) considered this problem in the scope

[1] At this stage of the considerations,the problem of spare loaders is ignored. This will be covered in chapter 12.

of the theory of queues, whereas Smith (1955) connected it with renewal theory. It has been stated that for a number of random processes such an event exists that if it appears the further course of the process does not depend on its past. Discussions concerning the regeneration of stochastic processes at first made only pure theoretical sense. Later, a practical meaning was found. Jastriebrienickij (1969) looked at alternative regenerative processes, and in this way developed the theory of regenerative processes. These processes may be applied in the investigation of the alternative renewal processes of systems operating in different regimes of work. Birolini (1971) discussed the problem of the application of regenerative processes to reliability theory. His discussion was based on definitions formulated by Smith[2] (1958). Borozdin and Yejov (1976) considered processes with strong regeneration, where a period of regeneration consists of two phases, one of exponential distribution and the second of general distribution. The publications of Rykov *et al.* (1971) and McDonald (1978) described processes with different kinds of regeneration. Berman (1978) discussed regeneration in point type processes, whereas Arndt and Franken (1979) considered the generalized regenerative processes.

It is worth pointing out that the second event of this type occurs when none of the trucks transport. The probability of the appearance of this state decreases with an increase in trucks employed in the system, and increases when the reliability of the hauling machines is reduced. The following is an identical situation—when trucks start their job again, the exploitation characteristics of the system are different to those for the stabilized period of exploitation.

A problem that is associated with the occurrence of a singular state during machinery system exploitation is the period of stabilization of the process. Czaplicki (2004, pp. 55 and 57) presented such periods after commencement of truck operation based on a deterministic simulation.

Knowing the normalization constant c (10.6), the conditional expected value of time loss in a truck work cycle due to truck standstill waiting for loading can be calculated. This mathematical hope is given by:

$$\Delta_d = cZ'P_{kd}^{(p)}\left(1-\sum_{g=0}^{d} p_{gd}\right). \tag{10.7}$$

Now the decisions of the truck dispatcher must be taken into account.

Here only the decisions connected with the reliability of the equipment involved, or in other words, associated with the processes of changes in reliability states of machines will be discussed.

Generally, truck dispatching can be defined as the current procedure of control of allocation of trucks. There are two grounds for this control:

- The implemented rule of truck dispatching, e.g. maximize output of shovels
- The current decisions of the truck dispatcher depending on the situation in the pit.

Additionally, both the above factors have lower significance in relation to the actual production requirements. The implemented rule can be changed if the nearest planned mine output must be modified, e.g. acceleration of removal of waste from the pit. Therefore, two factors influence truck dispatching:

- Exploitation states of the machines and shop, especially reliability ones
- Mineral production of the mine.

Although the above are not a great problem, the point is that the essential part of truck dispatching is prediction—forecasting the situation in the pit, the state of machines and their production in the near future. The majority of components being the basis of a decision in truck dispatching

---

[2] Smith (1958) gave the following definition: 'a stochastic process with a denumerable number of states z_0, z_1, z_2, … is said to be a regenerative process with respect to the state z_k if consecutive transitions into z_k (i.e. consecutive occurrence points of z_k) constitute a renewal process, the so-called imbedded renewal process'.

have a partly deterministic, partly stochastic nature. Additionally, the exploitation situation in the mine changes dynamically for many reasons: continuous enlargement of the pit, changing mining and geological conditions which are partly unknown, changes in weather and in the properties of machines and so forth. For these reasons, it is extremely difficult to find a dispatching rule with optimal parameters. At present, the only 'effective' tool in truck dispatching is simulation technique, but this too has some inconvenient properties.

It is now time to return to the main theme of these considerations.

It is necessary to consider cases when shovels change their states to a state of inability for loading, i.e. failure occurs in the shovel, or it is out of work for a lengthy duration for various reasons. First, the case that all shovels were loading and now one shovel is out should be considered. The dispatcher knows that he should withdraw some trucks to reserve. His decision is based on the present situation in the mine, predicted circumstances and his knowledge and experience.

For the system calculation, it is not important how many trucks will be extracted from the pit. Here it is assumed that the truck dispatcher keeps a constant ratio: the number of trucks directed to work to the number of shovels able to load. This principle will be marked by $\mathfrak{Z}$, i.e.:

$$\mathfrak{Z}: \frac{m_d}{d} = const \qquad \text{for } d = 1, 2, ..., n. \tag{10.8}$$

A different approach to this problem can be formulated based on the presented procedure. Based on formulas (8.2) and (8.4), all pairs $< m_d, r_d >$, $d = 1, 2, ..., n$ can be sought. It appears that pairs obtained in this way will be a little lower in value than pairs attained by applying the principle $\mathfrak{Z}$.

Now the values of the conditional parameters constructed can be analyzed, it is possible to verify the rationale of decisions made by the truck dispatcher. It makes significant practical sense. This is the basic scope of the estimation of the *goodness* of these decisions. Measures that should at first be taken into account are:

- The conditional probability p_{0d}
- The conditional measure of time loss Δ_d
- The conditional expected value E_{wkd}.

In some cases, more subtle and deep inferences should be made. Examples of such analysis will be presented in chapters 11 and 13, and section 14.2.

In order to continue the analysis it is necessary to consider:

- Pairs $< m_d, r_d >$, $d = 1, 2, ..., n$ when the grounds for constructing the probability distribution of the number of trucks in work state is the distribution obtained from the Maryanovitch model,
- Pairs $< m_d, r_d >$, $d = 1, 2, ..., n$ together with the number of repair stands k, when it is known that the probability distribution of the number of trucks in work state will be different than that from the Maryanovitch model.

When the number of shovels in a state of inability for loading increases and the number of trucks directed to accomplish the transportation task decreases, the relative number of surplus trucks increases. This number will be greater than that resulting from the cancellation of the effect of truck inaccessibility (it should be 15% from sequent 1.1 m_d). The outcome of this phenomenon is an increment in the truck reserve size.

If it is assumed that the principle $\mathfrak{Z}$ (10.8) holds, then the number of spare units increases by

$$\frac{0.15 \times 1.1\, m}{n} \qquad \text{for } n > 1 \tag{10.9}$$

each time, when the next shovel goes out of operation.

Practically, this increment will not be so great because the number of surplus units is also not great.

When both the number of loading shovels and the number of trucks working in the pit are decreasing, the service requirements (the number of failed trucks that need repair) also go down.

This is important for a system of trucks of low reliability. Following this line of consideration, when there is a shop with stands of low reliability each shovel failure is advantageous for the shop people. Moreover, the mean time that a truck spends at the shop waiting for repair also decreases. But it is known that in practice such a state does not last long. The mean time of shovel repair takes a good couple of hours. Besides, as a rule, mines have spare loading units; usually, front-end loaders. These machines resume loading duties when shovels fail. However, the mean time of loading by a wheel loader is much longer than that of a shovel. The loader has to move to put a load into the box of the truck; the shovel only revolves itself. The shovel opens a flap to release the loaded material; the loader must tilt its bucket. Generally, the effect of application of loaders when the shovel is in down state can be included in the procedure. This problem will be considered in chapters 12 and 13.

For now it is assumed that there are no spare loading machines in the system. If so, the organization of the system gradually changes when sequent shovels go down. As the number of trucks in the pit decreases, the number of trucks in reserve goes up. This means that the probability distribution of the number of trucks in work state becomes asymmetric, especially when the reliability of trucks is high. Sometimes it is enough when 2 or 3 shovels are unable to load and this probability distribution becomes point one with the probability $P^{(p)}_{wm_d} \cong 1$. A large reserve makes the system almost totally reliable—there is always a truck in reserve to replace the failed truck in the pit.

Figure 10.1 shows the probability distributions of a number of trucks in work state for two systems:

$$\mathfrak{S}_{\mathrm{XI}} : < m = 78,\ r = 16,\ k = m + r;\ A_w = 0.705 >$$

and

$$\mathfrak{S}_{\mathrm{XII}} : < m = 65,\ r = 29,\ k = m + r;\ A_w = 0.705 >,$$

i.e. after the withdrawal of 13 trucks to the reserve.

If computation is to be precise and it is taken into account that in this system there are 6 shovels (78/16) and the number of surplus units is 12 ($\cong 0.15\ m$), more trucks should be in reserve. In such a case the second probability function, marked in the dark colour, would be confined almost to this one high column.

Figure 10.2 is a similar diagram presenting the probability distributions of a number of trucks in work state for the same data but a different steady-state availability of truck, now $A_w = 0.790$.

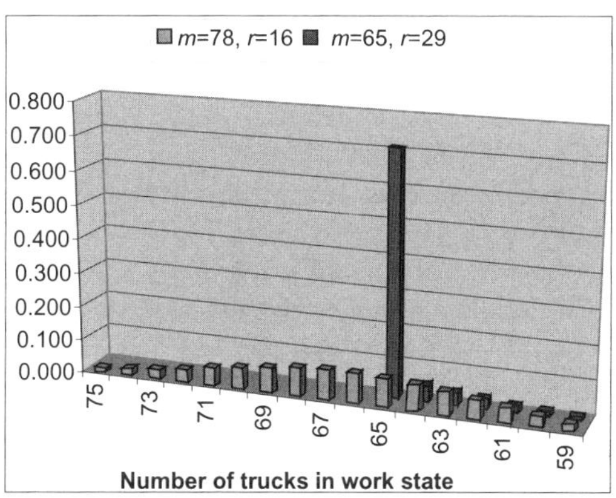

Figure 10.1. Probability distributions of numbers of trucks in work state for systems: $\mathfrak{S}_{\mathrm{XI}}$ and $\mathfrak{S}_{\mathrm{XII}}$ for the steady-state availability of truck $A_w = 0.705$.

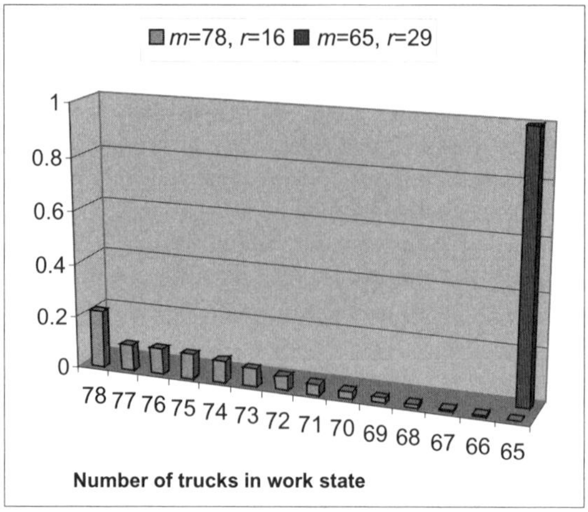

Figure 10.2. Probability distributions of numbers of trucks in work state for systems: $\mathfrak{S}_{XI}$ and $\mathfrak{S}_{XII}$ for the steady-state availability of truck $A_w = 0.790$ (notation of A)!

Normally, if the truck's reliability is higher, the number of spare units would be lower, but here it is assumed that there is no change, so that some regularities can be read easily.

Looking at Figure 10.2 in comparison to the previous diagram it is easy to see that, first, the mass of probability is shifted towards the maximum number of trucks that can be used in a pit. The second conclusion is that for a higher availability of trucks it is enough for one shovel to be down—the number of spares increases to such a degree that the system of trucks becomes totally reliable. There is always a certain truck in reserve to replace the truck that has just failed in the pit.

It is now time to return to the analysis and calculation.

It can be assumed that knowing the decisions made by the truck dispatcher and knowing what kind of procedure to apply, the probability distribution of the number of trucks in work state $P_w^{(p)}$ for each pair $< m_d, r_d >$ can be constructed. If so, the probability distribution of the number of trucks at shovels for each pair can be constructed, applying formulas (9.13) to (9.18). These will be the grounds for finding the conditional parameters (10.1) to (10.3) for all cases. Thus, the conditional approach can be left behind.

The probability that there will be no truck at shovels—all shovels waiting for a truck—is given by:

$$p_0 = c\sum_{d=1}^{n} p_{0d} P_{kd}^{(zd)}. \tag{10.10}$$

Similarly, the probability that there will only be one truck loaded at shovels, is given by:

$$p_1 = c\left((1 - p_{01}) P_{k1}^{(zd)} + \sum_{d=2}^{n} p_{1d} P_{kd}^{(zd)} \right). \tag{10.11}$$

The probability that there will be two trucks loaded is:

$$p_2 = c\left((1 - p_{02} - p_{12}) P_{k2}^{(zd)} + \sum_{d=3}^{n} p_{2d} P_{kd}^{(zd)} \right). \tag{10.12}$$

Further probabilities can be obtained in a similar way. Obviously, the number of simultaneously loaded trucks can be at most the same as the number of loading machines.

The above probabilities are essential parameters of the system efficiency of exploitation in the case of a truck subsystem. Properties of the shovel subsystem are also taken into account, but in an indirect form.

The following defines some important operation measures of the truck subsystem.

Having specified the probability distribution of a number of loaded trucks, the following expected values can be found:

- The unconditional mean number of loaded trucks at any moment of time:

$$E_{wlk} = \sum_{d=1}^{n} dp_d \tag{10.13}$$

- The unconditional expected number of trucks at shovels:

$$E_{wk} = c\sum_{d=1}^{n} E_{wkd} P_{kd}^{(zd)}. \tag{10.14}$$

It appears at first glance that these measures give the same information. This is not true. The following inequality holds:

$$E_{wk} > E_{wlk}$$

because the expected value determined by formula (10.13) is the sum of the trucks just loaded and trucks waiting for loading. The formula:

$$\theta = \frac{c}{n}\sum_{d=1}^{n} P_{kd}^{(zd)} E_{wkd} \tag{10.15}$$

describes the mean truck queue length per one shovel.

The mean number of trucks in work state in the system can be found using the formula:

$$E_w = c\sum_{d=1}^{n} E_w(m_d, r_d) P_{kd}^{(zd)}. \tag{10.16}$$

A very important parameter of the system—and the truck subsystem especially—is the mean time of the truck work cycle determined by

$$T_c = Z' + T_j + \sum_d \Delta_d. \tag{10.17}$$

An interesting measure too is the percentage increment of the mean time of truck work cycle due to trucks waiting for loading. This measure can be computed from the formula:

$$\frac{\sum_d \Delta_d}{Z' + T_j} 100. \tag{10.18}$$

The following is an evaluation of the system output.

The theoretical shovel system loading capacity counted in number of trucks loaded is:

$$W_{tk} = \frac{60}{Z'} n \qquad \text{trucks/h.} \tag{10.19}$$

if the truck work cycle components are given in minutes.

This is the shovel system output provided that the machines are totally reliable, continuously accessible and there is always a truck to be loaded.

If shovel reliability is taken into account, then a formula determining the potential output can be obtained as follows:

$$W_{pk} = \frac{60}{Z'} n A_k \qquad \text{trucks/h.} \tag{10.20}$$

If shovel accessibility is included in this consideration, then it will be the shovel system's own output:

$$W_{ok} = \frac{60}{Z'} n A_k B_k \qquad \text{trucks/h.} \qquad (10.21)$$

This production takes into account shovel accessibility and reliability but assumes that there is still a truck to be loaded.

The system output, also taking into consideration the fact that sometimes there will be no truck at shovel, is the shovel system effective output calculated from:

$$W_{efk} = \frac{60}{Z'} E_{wlk} \qquad \text{trucks/h.} \qquad (10.22)$$

This is the shovel production that can be expected in a longer period of time, provided that the system parameters do not change.

An interesting system parameter is the following ratio expressed in %:

$$\frac{W_{efk}}{W_{wk}} 100 = \frac{E_{wlk}}{nG_k} 100 = 100 \frac{\textit{mean number of trucks loaded in any moment of time}}{\textit{mean number of shovels able load}} \qquad (10.23)$$

This measure takes into account both losses, those connected with the lack of trucks to be loaded when shovels are waiting for trucks, and those associated with trucks waiting idly for loading. This ratio is in practice obviously below 100%. The greater the influence of the stochastic nature, the greater the dispersion of values of random variables of the exploitation process and equipment involved, the lower its value is.

The truck system output can be defined as:

$$W_{pw} = \frac{60}{Z' + T_j} E_w \qquad \text{truck work cycles/h.} \qquad (10.24)$$

If the truck system consists of m trucks directed to accomplish the transportation task, and the system also contains spare loading machines assuring that there will be a constant number of operating trucks, the expected value E_w is calculated from the Maryanovitch model. This mean number of trucks in work state is given by:

$$E_r(D) = p_0 \left\{ \sum_{k=0}^{r} \left(\frac{1 - A_w}{A_w} \right)^k \frac{m^{k+1}}{k!} + m^{r+1} \sum_{j=r+1}^{m+r-1} \frac{(1 - A_w)^j}{j! A_w} \prod_{l=r+1}^{j} (n + r - l) \right\} \qquad (10.25)$$

where p_0 is the probability that 0 trucks are in repair.

Formula (10.24) takes into account the reliability of both trucks and repair stands if the set of formulas (7.27) and (9.11) is applied. Moreover, it considers changes in the organization of the system as well as the accessibility of loading shovels. It does not discuss losses in time in the truck work cycle due to the existence of a truck queue at the shovel.

The truck effective output can be computed from the expression:

$$W_{efw} = \frac{60}{T_c} E_w \qquad \text{truck work cycles/h.} \qquad (10.26)$$

Time losses are included in this formula.

The following relationship should hold: $W_{efk} \cong W_{ef}$.

It can be assumed that the lower value of the above two: W_{efk} and W_{ef} is the effective machinery system output.

CHAPTER 11

Modelling—Case study I

In the previous chapters, many formulas as well as large parts of the calculation procedure have been considered—it is time now to show how this all works in practice. So in this chapter an example machinery system will be discussed.

It will be assumed that there is a machinery system in operation in a mine, i.e. there are n power shovels loading and, for the time being, this number will not change. The reliability of these machines is described by the steady-state availability A_k, and their accessibility is assessed as B_k. The bucket capacity and properties of the excavated rocks means that trucks have been selected according to their type $\mathfrak{T}$ and payload Q. Therefore the mean adjusted loading time Z' has been estimated and the corresponding standard deviation σ_z. Information on the reliability of haulers has been gathered, resulting in the estimation of the basic reliability parameters. Thus, the mean time of work state δ^{-1} and the corresponding standard deviation σ_p, the mean time of repair γ^{-1} and the corresponding standard deviation σ_n are known. The mean time of truck travel has also been estimated as T_j for all trucks and shovels and possible locations of working places. The corresponding standard deviation of truck travel has been assessed at σ_j.

The following data is assumed:

$$\mathfrak{S} : < \mathfrak{H}: n = 5,\ A_k = 0.82,\ B_k = 0.80;$$

$$\mathfrak{W}: \delta = 0.05\ \mathrm{h}^{-1} = (\sigma_p)^{-1},\ \gamma = 0.19\ \mathrm{h}^{-1},\ \sigma_n = 2.1\ \mathrm{h};$$

$$Z' = 2.1\ \mathrm{min};\ \sigma_z = 0.4\ \mathrm{min},\ T_j = 38\ \mathrm{min};\ \sigma_j = 17\ \mathrm{min} >$$

where:

$\mathfrak{H}$—system of power shovels,
$\mathfrak{W}$—system of trucks.

The goals of this considerations are:

1. Selection of a truck fleet size that is the selection of a pair $< m, r >$, m—number of dumpers directed to execute hauling duty and number r trucks that should be in reserve; determination of surplus haulers
2. Verification of whether the truck backup facility is enough; criterion: the mean time that a truck waits for repair should be negligible
3. Determination of the probability distribution of the number of trucks in work state
4. Determination of the probability distribution of the number of shovels in accessibility state
5. Evaluation of the conditional parameters of the shovel-truck system and verification of the quality of decisions made by the truck dispatcher
6. Evaluation of the unconditional parameters of the shovel-truck system
7. Productivity system assessment
8. Analysis of results, general conclusions and recommendations, remarks.[1]

[1] Point 8 contains general inferences and suggestions; specific, local ones will be presented during the analysis: points 4–7.

Ad. 1 The first task is the selection of the pair $< m, r >$.

The number of trucks required to accomplish the transportation task determined by the shovel system can be obtained from formula (8.2):

$$h = B_k A_k n \frac{Z' + O + W + R}{Z'} = 62.63\,.$$

Taking into account the steady-state availability of trucks, gives (8.4):

$$V = \frac{h}{A_w} = \frac{62.63}{0.792} \cong 79.1$$

The hauling system should possess 80 trucks, if their accessibility is ignored.

This system can be divided into trucks directed to execute a hauling job and the rest, which form the reserve. Sequent machines can be withdrawn to reserve and—applying formula (2.1) from the Maryanovitch model—a pair is sought to satisfy $< m, r \mid r_{max} >$ to obtain a number of trucks in work state not lower than h. After a short calculation, the pair is found:

$$< m = 65, r = 15 >.$$

If the number of trucks directed to the pit is increased by 10%, then:

$$< m = 72, r = 15 >.$$

The probability distribution of the number of failed trucks for the pair $< m = 72, r = 15 >$ is given in Figure 11.1.

Now consider how many trucks should be in the system.

Taking into account truck accessibility, the truck system should have 83 units because:

$1.15\, m \Rightarrow 83$ haulers that is: 72 trucks + 11 surplus units.

Later on in the analysis of the truck workshop, it is assumed that there are 83 machines.

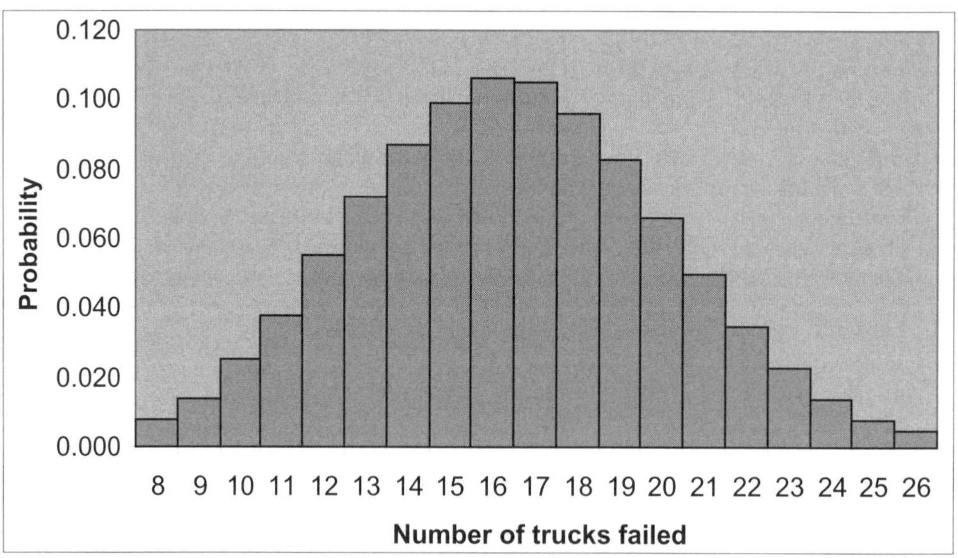

Figure 11.1. Probability distribution of the number of failed trucks for system of trucks $< m = 72, r = 15 >$ and truck steady-state availability $A_w = 0.792$.

Ad. 2 Number of repair stands required

Knowing that it is recommended to have a higher number of repair stands than number of machines $r > k$, it is necessary to find the appropriate number of repair stands for the truck system in the following range:

$$\frac{\delta}{\gamma} m = 21.8 \leq k.$$

Therefore, the following sequence of possible numbers of stands is considered:

$$k = 23, 24, 25, 26, 27.$$

Applying formulas (7.5), (7.9)–(7.13), the probability distribution of the number of failed trucks for a given number of repair stands is constructed. On this basis, four selected parameters are estimated employing appropriate formulas:

- The mean time of a truck waiting for repair T_{ow}; formula (7.15)
- Increase of the mean time of truck unserviceability ϕT_n; formula (7.16)
- The mean number of failed trucks E_u; formula (7.19)
- The adjusted steady-state availability A'_w; formula (7.17).

The outcomes of computations are given in Table 11.1.

The results of computations allow the following recommendation to be made:

It is advisable to have 27 repair stands in the system workshop; however, 1 or 2 repair stands less will not make a great difference to the operation parameters of the system.

It is also necessary to check whether the heavy traffic condition fulfilled in the repair shop is considered. Estimating the intensity flow rate coefficient, gives:

$$\frac{m\delta}{k\gamma} = 0.809.$$

It can therefore be said that the heavy traffic condition is being fulfilled.

For the further stage of the modelling, a system of the following structural parameters is assumed:

$$< m = 72, r = 15, k = 27 >.$$

The expected number of busy repair stands for the selected system can now be calculated. Applying formula (7.21), gives:

$$E_{BRS} = 20.4,$$

which gives the operative utilization of the repair shop as: $OU = 0.76$.

It can be concluded that the planned repair shop appears to be well selected in terms of number of repair stands.

Table 11.1. Parameters of investigated systems.

k	Repair stands	23	24	25	26	27
T_{ow}	h	3.4	1.6	0.7	0.3	0.1
ϕT_n	h	0.8	0.2	0.1	0.0	0.0
E_u	trucks	20.9	20.6	20.5	20.4	20.4
A'_w		0.765	0.784	0.790	0.791	0.792

Ad. 3 Construction of the probability distribution of the number of trucks in work state and computation of the mean number of trucks in work state.

Two pieces of information are significant for the final shape of the probability distribution of the number of trucks in work state, which is necessary for the next stage of the modelling. These are:

- Number of repair stands in relation to the number of trucks in the system, and
- Reliability of repair stands.

It is assumed that the number of repair stands in the workshop will be such that any queue of failed trucks before the shop will be insignificant. This is convenient both for mine practice and for further calculation. No losses in time for truck, no losses in time during modelling—no need to lump states and more complicated computations.

The reliability of repair stands is a separate problem, as was proved in chapter 9.1. If a specific mine has unreliable repair stands the probability distribution (7.27) should be constructed using formulas (9.11) and the appropriate patterns to go with these.

If the repair shop is well organized, possesses repair stands of high reliability and there are one or two spare stands, it can be assumed that the shop is almost totally reliable and thus apply the Maryanovitch model.

It is assumed that this mine has a very good, well-organized repair shop. For this reason, formulas (7.27) and (2.1) can be used.

The probability distribution of the number of trucks in work state is shown in Figure 11.2.

The expected number of trucks in work state $E_w(72,15) = 68.5$.

Ad. 4 Determination of the probability distribution of the number of shovels in accessibility state.

The system of shovels consists of 5 machines working independently of each other. In reliability terms, this is a system of five objects operating simultaneously in parallel. The reliability of this system is determined by distribution (7.1). Taking into account the accessibility of these machines, this distribution has to be modified to obtain the distribution described by formula (7.4).

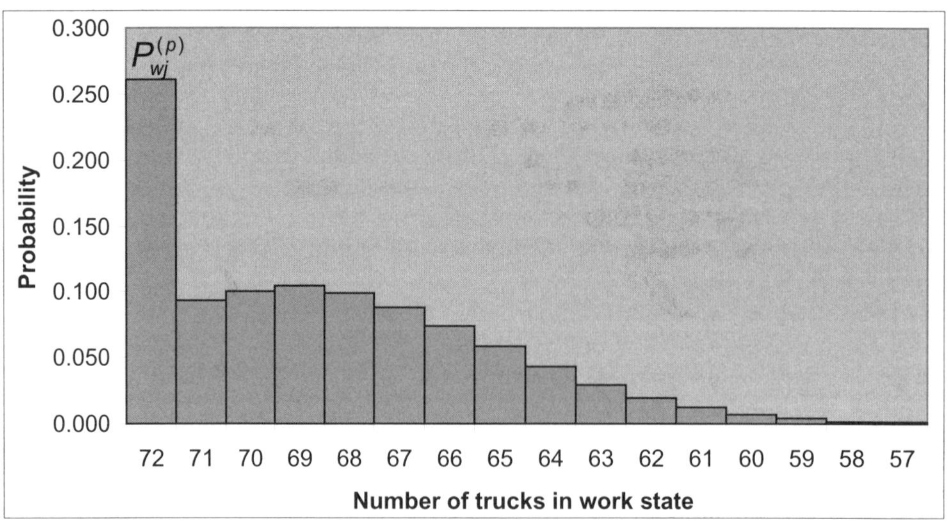

Figure 11.2. Probability distribution of number of trucks in work state for system < m = 72, r = 15 > and the steady-state availability $A_w = 0.792$.

For the data considered in this chapter, the probability distribution of the number of power shovels in a state of accessibility for loading (able to load) is shown in Figure 11.3.

Ad. 5 Evaluation of the conditional parameters of the shovel-truck system and verification of the goodness of decisions made by the truck dispatcher

Before starting to construct the most important characteristics of the machinery system at this stage of modelling and analysis, it is necessary to verify whether the heavy traffic condition is fulfilled, similarly to when considering a truck-workshop system.

The loading shovels now determine the service system and the clients are haulers. The heavy traffic situation is determined by condition (9.19), that is:

$$\frac{1.1\,m}{n}\frac{Z'}{T_j} \cong 0.796 \geq 0.75.$$

The condition is fulfilled.

Therefore, the probability distribution of the number x of trucks at shovels—formula (9.18)—cam be constructed. It is necessary to consider case-by-case when the number of shovels able to load equals $d = 5, 4, \ldots, 1$, using formulas (9.14a), (9.14b), (9.15a)–(9.15c), (9.16) and (9.17). This probability distribution can be used to find the set of conditional parameters of the system.

The following discussion considers this case-by-case.

- Starting from the system[2]:

$$\mathfrak{S}_5 : < n_5 = 5;\ m_5 = 72,\ r_5 = 15,\ k = 27 >$$

which means that 5 power shovels are able to load, 72 trucks are directed to haul, 15 trucks are in reserve and the repair shop has 27 repair stands. The probability of occurrence of this state is 0.121.

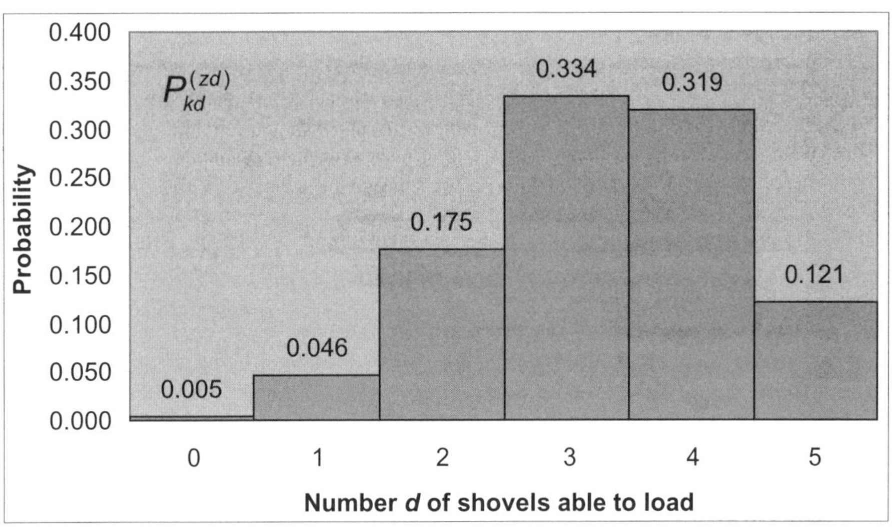

Figure 11.3. Probability distribution of number of shovels able to load.

[2] Some calculation procedures have several steps. The symbol • indicates the start of a new step.

Recalling the formula describing the probability density function of the number of trucks at shovels:

for $0 \le x \le d = 5$

$$K_{55} f_5(x;d,b) = \frac{K_{55}}{(b-x)\varpi C_1 + xC_2} \left(\frac{(b-x)\varpi C_1 + xC_2}{b\varpi C_1} \right)^{b\beta_5'} \exp\left(\frac{2(\varpi + 1)x}{\varpi C_1 - C_2} \right)$$

because $\kappa C_1 - C_2 \neq 0$

and for $5 \le x \le b = 72$

$$K_{66} f_6(x;d,b) = \frac{K_{66}}{(b-x)\varpi C_1 + dC_2} \left(\frac{(b-x)\varpi C_1 + dC_2}{(b-d)\varpi C_1 + dC_2} \right)^{d\beta_3'} \exp\left(\frac{2(x-d)}{C_1} \right).$$

Employing formulas (9.16) and (9.17), the basic conditional parameters can be found. The conditional probability that there will be no trucks at shovels is:

$$p_{0d=5} = \sum_{b}^{72} P_{wb}^{(p)} K_{d55}(b) \int_0^{0.5} f_5(x;5,b)dx = 1 \times 10^{-7}$$

Further probabilities are as follows:

$p_{15} = 0.000$ $p_{25} = 0.046$ $p_{35} = 0.401$ $p_{45} = 0.465$ $p_{55} = 0.082$ $p_{>55} = 0.006.$

The conditional expected number of trucks at 5 power shovels able to load is:

$$E_{wk5} = \sum_{b}^{72} P_{wb}^{(p)} K_{566}(b) \left(\Psi_5(b) \int_0^5 x f_5(x;5,b)dx + \int_5^b x f_6(x;5,b)dx \right) = 3.6 \text{ trucks}$$

and the conditional truck mean time loss in the truck cycle due to waiting in a queue for loading is given by:

$$\Delta_5 = cZ' P_{k5}^{(zd)} \left(1 - \sum_{g=0}^{5} p_{g5} \right) = 0.0 \text{ min}.$$

Thus, a *local* level, the most frequent cases are of 3 or 4 dumpers being at shovels. This means that 2 or 1 power shovels wait idly for trucks.

The conclusion for the truck dispatcher:

When all power shovels are able to work—that is in 12% of cases—direct more trucks to the pit because the shovel system is not heavily loaded from the service point of view.

Notice that, strengthening this interference, on average there are 3.6 trucks at 5 power shovels able to load.

- Now consider a case where one shovel is down, failure occurs. There are no spare loaders, so the dispatcher will withdraw some trucks from circulation. If it is assumed that the principle **3** is valid the dispatcher intends to withdraw 14 trucks or, better, a few haulers less, bearing in mind the previous situation. So, it is assumed that 10 trucks have been directed to reserve. It is then necessary to immediately take into account that some of the surplus trucks—1/5 from 11—should be added to the reserve. Thus the system taken into consideration will be:

$$\mathfrak{S}_4 : < n_4 = 4;\ m_4 = 62,\ r_4 = 27,\ k = 27 >.$$

The probability that such a system should be considered is 0.319.

The Maryanovitch model is now applied to get the probability distribution of the number of trucks in work state. This distribution appears as follows:

$$P_{w62}^{(p)} = 0.995 \quad P_{w61}^{(p)} = 0.002 \quad P_{w60}^{(p)} = 0.001 \quad P_{w59}^{(p)} = 0.001$$

and the expected number of trucks in work state: $E_w(62, 27) \cong 62.$

The probability density function of the number of trucks at 4 shovels able to load can now be built. The mathematical construction is identical to that presented previously.

The conditional parameters are calculated.

The conditional probability that there will be no trucks at shovels is:

$$p_{0d=4} = \sum_{b}^{62} P_{wb}^{(p)} K_{d55}(b) \int_{0}^{0,5} f_5(x;4,b)dx = 1 \times 10^{-6}$$

Further probabilities are as follow:

$p_{14} = 0.002$ $\quad p_{24} = 0.102$ $\quad p_{34} = 0.537$ $\quad p_{44} = 0.311$ $\quad p_{>44} = 0.048.$

The conditional expected number of trucks at 4 power shovels able to load is:

$$E_{wk4} = \sum_{b}^{62} P_{wb}^{(p)} K_{566}(b) \left(\Psi_5(b) \int_{0}^{4} x f_5(x;4,b)dx + \int_{5}^{4} x f_6(x;4,b)dx \right) = 3.3 \text{ trucks}$$

and the conditional truck mean time loss in the truck cycle due to waiting in a queue for loading is given by:

$$\Delta_4 = cZ' \, P_{k4}^{(zd)} \left(1 - \sum_{g=0}^{4} p_{g5} \right) = 0.0 \text{ min.}$$

A short comment should be made here. The dispatcher decided to increase the relative service load of the loading machines. The operational parameters changed advantageously. In the case actually considered, there are 3.3 haulers per 4 shovels able to load, on average. Additionally, the reserve size is large, thus there is no problem sending more trucks to the pit if the dispatcher decides on this course of action.

- The next case to discuss is when 2 shovels are in failure. The probability of this event was estimated as being the most frequent—0.334. Some further trucks can be withdrawn from circulation. It is assumed that 12 trucks have been directed to reserve. There is the following machinery system:

$$\mathfrak{S}_3 : < n_3 = 3;\ m_3 = 52,\ r_3 = 37,\ k = 27 >.$$

Observe that for a large reserve 2 or 3 trucks more or less do not make a difference. The number of spare trucks is so great that it can be assumed that 52 circulating trucks are totally reliable.

Proof of this statement can be obtained by looking at the results of the Maryanovitch model calculation.

The probability $P_{w52}^{(p)} \cong 1$ and the expected number of trucks in work state is $E_w(52, 37) \cong 52$.

For such a reduced probability distribution, the probability density functions of the number of trucks at 3 power shovels able to load are as follows:

for $\quad 0 \leq x \leq 3$

$$f_5(x;3,52) = \frac{1}{(52-x)\varpi C_1 + xC_2} \left(\frac{(52-x)\varpi C_1 + xC_2}{52\varpi C_1} \right)^{52\beta_5'} \exp\left(\frac{2(\varpi+1)x}{\varpi C_1 - C_2} \right)$$

for $\quad 3 \leq x \leq 52$

$$f_6(x;3,52) = \frac{1}{(52-x)\varpi C_1 + 3C_2} \left(\frac{(52-x)\varpi C_1 + 3C_2}{(52-3)\varpi C_1 + 3C_2} \right)^{3\beta_5'} \exp\left(\frac{2(x-3)}{C_1} \right).$$

The conditional probabilities appear to be:

$p_{03} = 0.000 \qquad p_{13} = 0.010 \qquad p_{23} = 0.256 \qquad p_{33} = 0.463 \qquad p_{>33} = 0.271.$

The conditional expected number of trucks at 3 power shovels able to load is:

$$E_{wk3} = K_{366}(52)\left(\Psi_5(52)\int_0^3 xf_5(x;3,52)dx + \int_3^{52} xf_6(x;3,52)dx \right) = 3.1.$$

and the conditional truck mean time loss in the truck cycle due to waiting in a queue for loading is given by:

$$\Delta_3 = cZ' P_{k3}^{(zd)} \left(1 - \sum_{g=0}^{3} p_{g3} \right) = 0.2 \text{ min}.$$

The exploitation situation has changed compared to the previous one. The probability that there will be 3 or more trucks at 3 shovels able to load is high: 0.734. For this reason, the average number of trucks at shovels is above the number of loading machines able to load. However, time losses in the truck work cycle are inevitable. The parameter $\Delta_3 \cong 0.2$ min.

- The penultimate case is a state where 3 shovels have failed. The probability of such an event is 0.175.

It is assumed that the truck dispatcher withdraws 12 further dumpers.

The machinery system is:

$$\mathfrak{S}_2 : < n_2 = 2;\ m_2 = 40,\ r_2 = 49,\ k = 27 >.$$

Again, the reserve size does not matter. It is very large.

The result of application of the Maryanovitch model is easy to predict.

The probability density function is similar to before; the differences are in the values of appropriate parameters.

The conditional parameters are as follows:

$p_{02} = 0.000$ $p_{12} = 0.019$ $p_{22} = 0.137$ $p_{>22} = 0.844$

and the conditional expected number of trucks at 2 power shovels able to load is:

$$E_{wk2} = K_{366}(40)\left(\Psi_5(40)\int_0^2 xf_5(x;2,40)dx + \int_2^{40} xf_6(x;2,40)dx\right) = 4.3.$$

The conditional truck mean time loss in truck cycle due to waiting in a queue for loading is given by:

$$\Delta_2 = cZ'\,P_{k2}^{(zd)}\left(1 - \sum_{g=0}^{2} p_{g2}\right) = 0.3\ \text{min}.$$

Observing the value of the above parameters, it appears that the dispatcher's decision was not so good. In this state, an almost permanent queue of trucks is observed. Trucks will waste time standing in queues. However, if it is assumed that the best solution is to keep shovels almost always busy, here there is such a situation.

- The last situation to discuss is a state where 4 power shovels are down and only one loading machine is able to execute its duties. This is a rare event and its probability was assessed at 0.046.

  It is assumed that the dispatcher withdraws a lot of haulers, say 20 trucks.

  The machinery system to be discussed is:

$$\mathfrak{S}_1 : < n_1 = 1;\ m_1 = 20,\ r_1 = 69,\ k = 27 >.$$

Construction of the probability density function of the number of trucks at this 1 shovel able to load does not cause any problems.

The conditional parameters in this case are as follows:

$p_{01} = 0.012$ $p_{11} = 0.236$ $p_{>11} = 0.752$

and the conditional expected number of trucks at the 1 power shovel able to load is:

$$E_{wk1} = K_{366}(20)\left(\Psi_5(20)\int_0^1 xf_5(x;1,20)dx + \int_1^{20} xf_6(x;1,20)dx\right) = 2.4.$$

The conditional truck mean time loss in the truck cycle due to waiting in a queue for loading is given by:

$$\Delta_1 = cZ'\,P_{k1}^{(zd)}\left(1 - \sum_{g=0}^{1} p_{g2}\right) = 0.1\ \text{min}.$$

Some remarks are again advisable. Instead of withdrawing so many trucks, the operation situation in the pit is good in terms of the production of the mine. For almost 99% of the time a shovel will have at least one truck to load. The average truck queue is 1.4—not so long. Similarly, the mean time loss is not great, only about 0.1 min on average.

Table 11.2 gathers and presents all the results of calculations.

In spite of the fact that the values of these parameters depend on the subjective decisions of the truck dispatcher, some conclusions can be formulated:

a. Observing the first column, where the probabilities of a shovel's standstill are given, it should be stated that the level of this probability is correct (shovels should work round-the-clock). The increase of this probability in the last case when only one shovel is able to load can be accepted for at least two reasons: the probability of this state is very low (0.046) and the values of parameters connected with the truck queue (column 8 and 10) indicate that there was no sense in increasing the number of circulating trucks.
b. Column 8 presents the probability of the occurrence of a queue of trucks for a given machinery system. It increases when sequent shovels go down, meaning that more trucks could be withdrawn from the pit. Only in the last case does this probability drop compared to the previous case. This decrement was made by withdrawing many trucks from the pit. If it is considered that shovels should work almost continuously, this decision is not advisable.
c. Column 10 shows the value of the measure of relative service load of the shovel system. This service load increases when the number of shovels able to load decreases. This regularity is observed in mine practice. Generally, it is good if this relative service load is high, provided that the value of loss parameter is low.
d. Analyzing values given in the last column, it can be seen that they look stable apart from the case of the system $\mathfrak{S}_2$. A significant increase in the mean truck time loss is observed due to there being many trucks in the pit.

Ad. 6 Evaluation of the unconditional parameters of the shovel-truck system

The whole shovel system will be stopped due to a lack of trucks with probability:

$$p_0 = c\sum_{d=1}^{n} p_{0d}P_{kd}^{(zd)} = 6\times 10^{-4}.$$

The probability that shovels will load only one truck is:

$$p_1 = c\left((1-p_{01})P_{k1}^{(zd)} + \sum_{d=2}^{n} p_{1d}P_{kd}^{(zd)} \right) = 0.053.$$

The probability that shovels will load 2 trucks is:

$$p_2 = c\left((1-p_{02}-p_{12})P_{k2}^{(zd)} + \sum_{d=3}^{n} p_{2d}P_{kd}^{(zd)} \right) = 0.297.$$

Table 11.2. Conditional parameters of analyzed system.

System	p_0	p_1	p_2	p_3	p_4	p_5	$p_>$	E_{wkd}	E_{wkd}/d	Δ
								trucks	trucks/shovel	min
1	2	3	4	5	6	7	8	9	10	11
$\mathfrak{S}_5$	*0.000*	*0.000*	*0.046*	*0.401*	*0.465*	*0.082*	*0.006*	*3.6*	*0.72*	*0.01*
$\mathfrak{S}_4$	*0.000*	*0.002*	*0.102*	*0.537*	*0.311*		*0.048*	*3.3*	*0.825*	*0.03*
$\mathfrak{S}_3$	*0.000*	*0.010*	*0.256*	*0.463*			*0.271*	*3.1*	*1.03*	*0.19*
$\mathfrak{S}_2$	*0.000*	*0.019*	*0.137*				*0.844*	*4.3*	*2.15*	*0.31*
$\mathfrak{S}_1$	*0.012*	*0.236*					*0.752*	*2.4*	*2.4*	*0.07*

The most frequent case is that 3 trucks will be loaded. Its probability is:

$$p_3 = 0.476.$$

Further probabilities are: $p_4 = 0.172$ $\quad$ $p_5 = 0.011.$

The probability distribution of the number of trucks loaded by power shovels for the analyzed system is illustrated in Figure 11.4.

Looking at this probability distribution, it appears to be a symmetric distribution, or with a light positive skew.[3] This suggestion can be supported by computing the value of the statistical measure of the skew. Selecting probably one of the most frequently used. Denoting by μ the third moment about the mean and by σ the standard deviation, the ratio:

$$\frac{\mu}{\sigma^3}$$

is the measure of skew (for example, Stuart and Ord 1998). In this case, it is 0.149. The outcome gives the information that this probability distribution has a light positive asymmetry. However, it is much better if this type of probability distribution has a light negative skew, indicating that the shovel system is heavily loaded from the service point of view. Observe that if this type of probability distribution possesses a strong skew this means that the system was not properly selected. Generally, the skew of distribution is a certain measure of how properly the elements of the machinery system were chosen and how they match each other.

The expected number of trucks loaded[4] is:

$$E_{wlk} = \sum_{d=1}^{n} dp_d = 2.82 \text{ trucks}$$

with a standard deviation equalling 0.83.

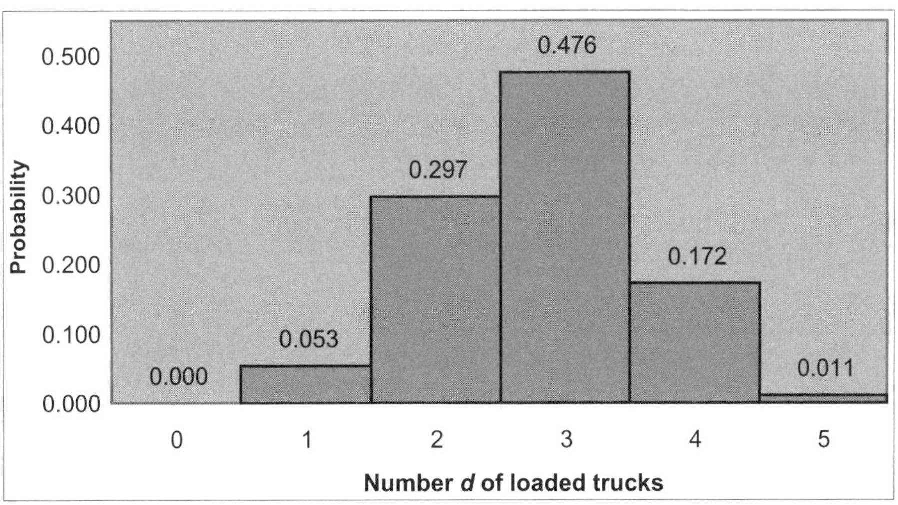

Figure 11.4. Probability distribution of number of loaded trucks by shovel system.

[3] The distribution is said to be right-skewed or with a positive skew if the mass of probability is concentrated on the left of the figure.

[4] This mean is also the conditional one—provided that there is a truck to be loaded.

The unconditional expected number of trucks in the shovel system is:

$$E_{wk} = c\sum_{d=1}^{n} E_{wkd} P_{kd}^{(zd)} = 3.40 \text{ trucks.}$$

Thus the mean number of trucks waiting for loading—the average truck queue length—in the shovel system is:

$$\theta = \frac{1}{n}\sum_{d=1}^{n} P_{kd}^{(zd)} E_{wkd} = 1.17 \text{ trucks.}$$

which is not so long.

The average number of trucks in work state in the system is:

$$E_w = c\sum_{d=1}^{n} E_w(m_d, r_d) P_{kd}^{(zd)} = 53.6 \text{ trucks.}$$

Here a *local* exploitation measure can be constructed, namely:

$\frac{E_w}{nA_k} = 16.3$ mean number of trucks in work state/1 shovel in work state.

At this stage of the mine development it can be said that for the specified machinery system there are 16.3 trucks in work state per 1 shovel in work state, on average.

The mean time of the truck cycle is:

$$T_c = Z' + T_j + \Sigma\Delta_d = 40.7 \text{ min}.$$

The increase of the mean time of the truck work cycle due to time lost in a queue is:

$$\frac{\sum_d \Delta_d}{Z' + T_j} 100 = 1.5\%.$$

Ad. 7 Productivity system assessment.

The theoretical shovel system loading capacity:

$$W_{tk} = \frac{60}{Z'} n \cong 143 \qquad \text{loaded trucks/h.}$$

The potential shovel system output:

$$W_{tk} = \frac{60}{Z'} nA_k \cong 117 \qquad \text{loaded trucks/h.}$$

The shovel system's own output:

$$W_{tk} = \frac{60}{Z'} nA_k B_k \cong 94 \qquad \text{loaded trucks/h.}$$

If the theoretical output is compared with the own system output, the difference visible is connected with the shovel loading capability coefficient G_k. This coefficient, expressed as a percentage, gives the information as to what extent the theoretical capacity of the shovel system could be utilized to load haulers.

The shovel system effective output:

$$W_{efk} = \frac{60}{Z'} E_{wlk} \cong 80 \qquad \text{loaded trucks/h.}$$

The following is a discussion of the ratio:

$$\frac{W_{efk}}{W_{wk}} 100 = 85\%.$$

The system of trucks was properly selected. However, three main factors:

- The decisions of the truck dispatcher at different stages in the system operation
- Time losses of trucks in queues waiting for loading
- The stochastic nature of the exploitation process of the machinery system

mean that the loading capacity of the power shovels applied is in 85% utilized. More careful analysis of the obtained figures can give an answer to what extent a given factor has an influence on this result.

The following is a discussion of the productivity of the truck system.

The truck system output:

$$W_{pw} = \frac{60}{Z' + T_j} E_w \cong 80 \qquad \text{truck cycles/h.}$$

This is the productivity of the hauler system if there is no queue of trucks waiting for loading.

Taking time losses in the truck work cycle into account, the truck system effective output is obtained:

$$W_{efw} = \frac{60}{T_c} E_w = 79 \qquad \text{truck cycles/h.}$$

It can be assumed that this figure is the effective output of the machinery system considered.

Ad. 8 Analysis of results, general conclusions and recommendations, remarks.

Many points have already been made in terms of the analysis of results during the consideration of particular points of the procedure—usually, just after a new interesting outcome has occurred and at the end of each step made. Thus, this conclusive analysis should be short and general.

The decisions of the truck dispatcher have a great influence on the operation/exploitation of the entire machinery system. This is a truism. Now, there are mathematical tools to estimate how good particular decisions are. Immediately following this assessment, conclusions were formulated, and later recommendations were made as to what should be done in the near future.

It is now possible to judge the quality of selection of the number of trucks applied to accomplish the transportation task determined by the production of power shovels. It is also possible to estimate how correct the selection of the number of trucks in reserve was. However, it is now known that the number of trucks in reserve is a random variable. This variable was not directly defined; it was not necessary to do so. Other stochastic processes determined it. It has also been noticed that this random number is important when all loading machines fulfil their duties, and the significance of the reserve size is reduced when the number of shovels able to load increases.

During the selection of the number of trucks it was decided that in the case considered, 72 trucks should circulate in the pit, 11 trucks will be surplus and 15 haulers will be spare ones. Therefore the total will be 98 trucks for 5 power shovels for a given stage of mine development. The reliability of machines, their operation parameters and the decisions of the truck dispatcher meant that an average of 16.3 totally reliable trucks will serve 1 totally reliable shovel.

Some further remarks can be made in analyzing the results of this procedure. The Reader can do this alone.

It is important to observe that these considerations, analysis and calculations concern the particular stage of mine development expressed here by the mean time travelled by trucks. Mining works are in progress, mines enlarge continuously and hauling distances constantly become longer. So an enlargement of the truck fleet should follow these changes. This problem will be discussed in chapter 15.

CHAPTER 12

Spare loaders

Mines with a shovel-truck system usually have, for various purposes, a number of front-end loaders (FEL). These loaders are employed when one or more loading shovels fail. Transport means only give a profit if they are in motion and are loaded. The planned sequence of mine development slows down when some loading machines fail. For this reason, spare loaders are a good solution. Theoretically, a spare loader could be just an additional power shovel. But in practice, this does not work. Even if the problem of the financial losses incurred keeping a spare shovel at a standstill is ignored, the problem of reaching the place where this machine is needed in a relatively quick time means that it is out of the question. A shovel in a hurry runs at a few km/h, especially in a pit. Apart from this, shovels are not designed to move long distances. Sometimes, if such exceptional journey happens, and a machine moves several kilometres, much will be made of this event, and large photos in technical literature are a certainty.

Wheel loaders are therefore much better. As well as moving much more quickly these machines are usually not so far away, sometimes nearby executing clean-up duties for instance, keeping in good condition working in the room around a shovel. It can be assumed without much error that they may resume their loading duties as spare loaders almost instantly.

The main problem that should be taken into account is an assessment of the mean time Z_l that FEL spends loading a truck. If this is compared to the mean Z' the following formula can be constructed:

$$Z_l = \tau Z' \tag{12.1}$$

where τ is the proportional coefficient and it can be assumed that $\tau > 1$.

Extension of the mean loading time has a significant influence on the number of trucks needed for the system to accomplish the given transportation task. Based on formula (8.2), Figure 12.1 shows

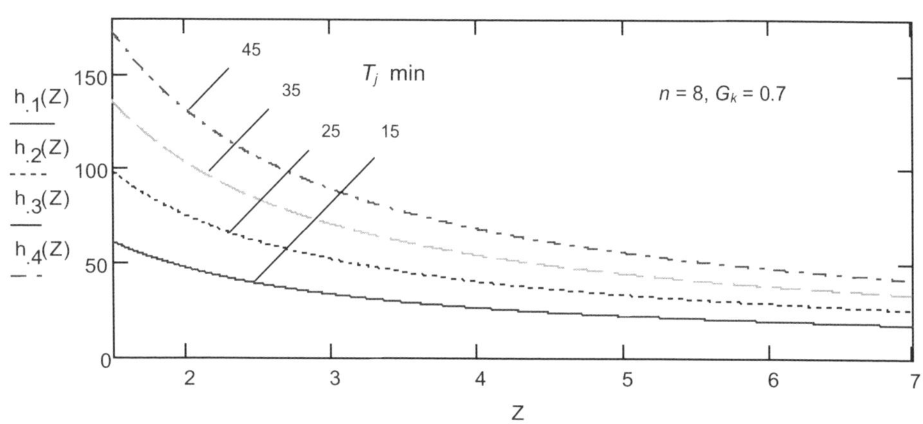

Figure 12.1. Example relationships between the number of trucks needed h and the function of mean loading time Z.

an example relationship between the mean loading time and the number of trucks needed for different mean truck travel T_j, ignoring the problem of truck reliability.

The coefficient τ is a function of two components. One factor that has an influence on its value is the bucket size of the loader compared to the size of the shovel bucket. As a rule, the bucket capacity of a front-end loader is not greater than the capacity of a bucket that has a shovel. Usually a bucket loader is smaller. The next factor is how *quick* a loader is compared to a shovel. A shovel is quicker at loading. During loading, a shovel does not displace itself, but only revolves, and a boom and stick are in motion together with some other parts and assemblies. Also unloading is simpler and quicker compared to the same action made by a wheel loader. According to technical literature (for example, Thomas 1979), if both loading machines with the same bucket/dipper capacity are compared, the output of a shovel will be approximately double that of a wheel loader.

If so, the mean time of loading for the system of n shovels if n_f shovels are down ($n_f \leq n$) and all failed machines are replaced by the same wheel loaders can be calculated from the formula:

$$Z'_n = Z'\left(1 + (\tau - 1)\frac{n_f}{n}\right). \tag{12.2}$$

In some places in this procedure more information was required than the mean loading time. The standard deviation of loading times is one thing that is needed. There are two ways to estimate this parameter. The best solution is an estimate based on empirical data, but this estimate is only precise for the machinery system in question. The second solution, of much lower likelihood and not needing any additional data, is the assumption that the coefficient of variation remains approximately constant. This means that if the standard deviation of loading times of the shovel is roughly 1/3 of the mean value, then this ratio can be maintained. If the real value is a bit higher or lower than that, the error generated is rather small.

Now it is necessary to take into consideration the reliability of these spare units. If these two types of loading machines are compaed, shovels are usually of greater reliability. The main point here is that shovels are designed in such way that they possess several different types of reliability redundancies in their construction. The performance of these machines is improved as far as reliability is concerned. They are strong machines well prepared to cope with difficult working conditions. Wheel loaders usually have slightly lower availability. But it can be assumed that when failure occurs in a wheel loader there is still an additional unit to replace this failed machine. For that reason, it can be assumed that spare units are totally reliable. Making such an assumption means that the reliability of the loading system is determined by the reliability of shovels exclusively, and formula (7.4) holds.

Now it is necessary to analyze how the calculation will change with wheel loaders always replacing failing shovels. First, it is necessary to specify what is going to change. Three system parameters will be different, namely:

a. The mean loading time of the truck; this parameter is now obtained from formula (12.2)
b. The standard deviation of loading times; a constant coefficient of variation can be assumed
c. The number of loading machines that is now constant and corresponding probabilities determined by the reliability of the shovel system.

An effect of the above changes will be clearly visible in the values of the system efficiency measures. In the whole set of formulas (9.12)–(9.18) one parameter will be different, that is ϖ given by formula (8.3). The second parameter with the mean loading time C_2 probably remains intact because it is the coefficient of variation.[1]

The main difference will be in the further part of the calculation procedure. The sequent cases when the number of shovels is going down does not cause any difference in computation apart

[1] If investigations in a given mine field give information that the empirical estimate of the coefficient of variation changes considerably, it will be easy to insert this new value to the procedure.

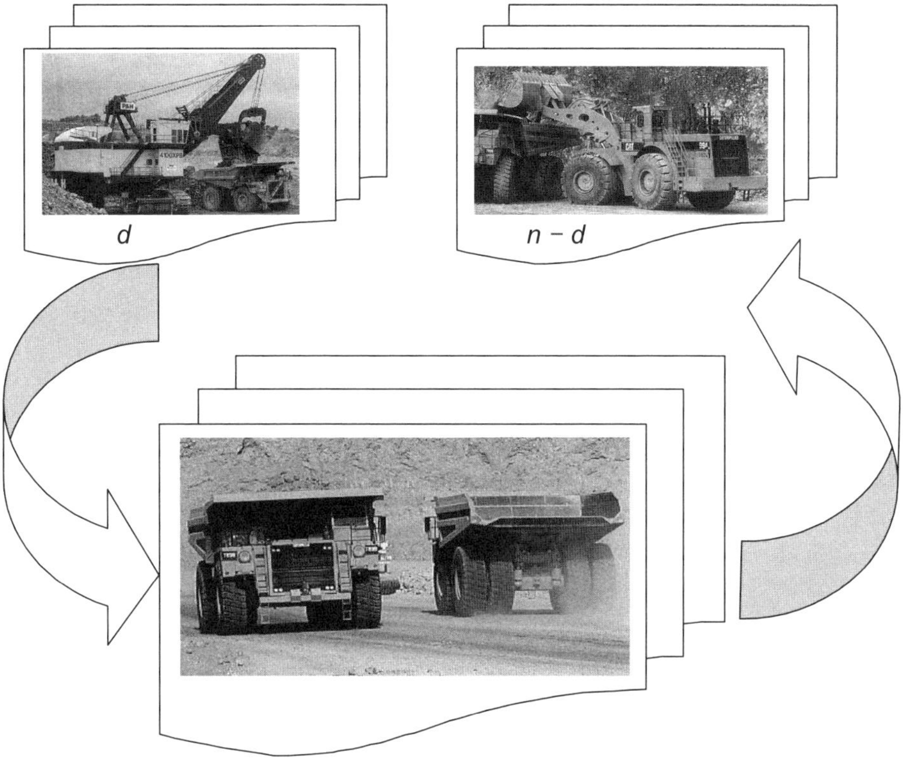

Figure 12.2. Idea of an enlarged loading system cooperating with trucks.

from an increment in the mean loading time. The number of loading machines is constant, and, compared to the previous calculation procedure, one randomization is less.[2] The calculation is repeated with the same number of hauling trucks, but in some cases it will not work.

It should be seen immediately that the decisions of the truck dispatcher will be different to those considered previously. If there is no spare loading machine, the truck dispatcher withdraws a certain number of trucks when the shovel is down. If the loading action is not stopped, and becomes a bit slower and longer lasting, it is not necessary to withdraw the trucks. But what will happen if the next shovels fail before the repair is finished in the first shovel? Generally, the mean time of the truck work cycle increases considerably. If so, it causes the mean number of trucks waiting for loading to increase too. Therefore, if the system of shovels consists of several units and the majority of them are down, the majority of trucks will soon be in queues waiting for loading. This increases the probability that a certain queue will block the transport road. Apart from that, there is no sense in having many haulers in queues. The truck dispatcher will withdraw some trucks to reserve. The organization of the system will change, and its calculation method must be modified. These statements can be supported by analyzing an example system.

[2] Notice that the constant $c = 1$ (formula 10.6).

CHAPTER 13

Modelling—Case study II

The following machinery system will be considered:

$$\mathfrak{S}_{XIV} : < \mathfrak{H}: n = 7, A_k = 0.84, B_k = 0.82;$$

$$\mathfrak{W}: A_w = 0.648, \delta = 0.077, \sigma_p = \delta^{-1}, \sigma_n = 0.35\ \gamma^{-1};$$

$$Z' = 2.2 \text{ min}, \sigma_z = 0.8 \text{ min}, T_j = 18.5 \text{ min}, \sigma_j = 8.7 \text{ min} >.$$

First, it is noticed that trucks in the system are of low reliability. Now the truck system and the number of repair stands for the repair shop can be selected.

Applying formulas (8.1)–(8.5), gives:

$$< 1.1\ m, r > = < 54, 22 >; 1.265\ m \Rightarrow 62.$$

For the above truck system, as many as $k = 37$ repair stands are needed to get a negligible queue of failed trucks waiting at the repair shop—formulas (7.9) to (7.13) with support of (7.5), (7.16) and (7.17).

The next step is the calculation of the probability distribution of the number of trucks in work state—formula (7.27). Employing the Maryanovitch model, the probability distribution shown in Figure 13.1 is obtained.

The expected number of trucks in work state for system $\mathfrak{S}_{XIV}$ is (formula (10.25)):

$$E(54, 22) \cong 49.1.$$

If the shovel system is now considered.

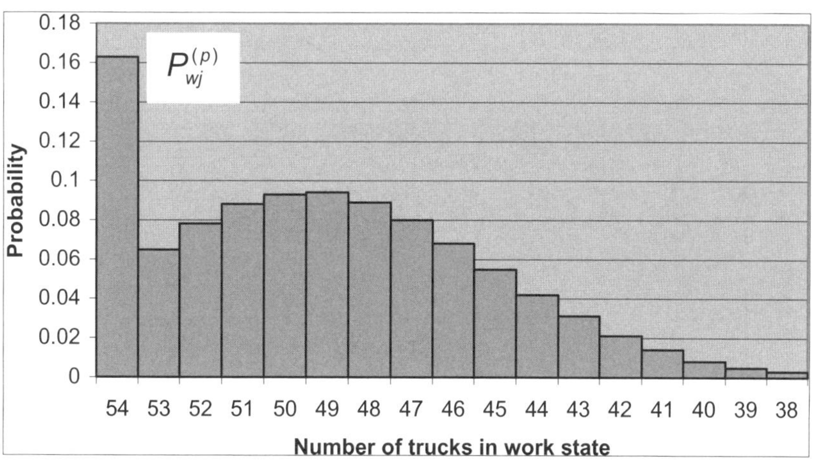

Figure 13.1. Probability distribution of number of trucks in work state for system $\mathfrak{S}_{XIV}$.

Using formula (7.4), the probability distribution of the number of shovels able to load is obtained. This is presented in Figure 13.2.

Before moving on to the next part of the modelling and analysis, it is necessary to specify whether the mine is equipped with spare loaders. If it assumed that front-end loaders are easily available, then the probability distribution $P_{kd}^{(p)}$, $d = 0, 1, \ldots, n$ shown on Figure 12.4, should be read in a slightly different way. The probability of an event that, say, 4 shovels are able to load means in fact that 4 shovels and 3 front-end loaders are able to load. Remember that a constant number of loading machines was assumed. For this reason, the notation is replaced with a new one:

$$P_{kd}^{(p)} \equiv P_{k[d+(n-d)]}^{(p)} \tag{13.1}$$

where the second part of the subscript should be read as: d shovels able to load and $(n - d)$ front-end loaders.

It is now necessary to specify the proportional coefficient τ. It can be assumed that after field investigation τ it was assessed to be $\cong 3$.

The modelling can now be started.

The above data is inserted into formulas: (9.12), (9.14a), (9.14b), (9.15)–(9.18).

The basic system parameters can now be calculated, but all cases connected with the reliability of the loading system have to be considered, one by one.

- The starting point is the case where all shovels are in work state as well as in a state of accessibility for loading, shortly able to load. The probability of this event is 0.074.
  The conditional probability that there will be a total lack of trucks at shovels is:

$$p_{0(7+0)} = \sum_{b}^{m_d} P_{wb}^{(p)} K_{d55}(b) \int_0^{0.5} f_5(x; d = 7, b)dx = 0.000.$$

Further conditional probabilities are:

$p_{1(7+0)} = 0.000$ $\quad p_{2(7+0)} = 0.002$ $\quad p_{3(7+0)} = 0.035$ $\quad p_{4(7+0)} = 0.200$

$p_{5(7+0)} = 0.378$ $\quad p_{6(7+0)} = 0.276$ $\quad p_{7(7+0)} = 0.087$ $\quad p_{>7(7+0)} = 0.021.$

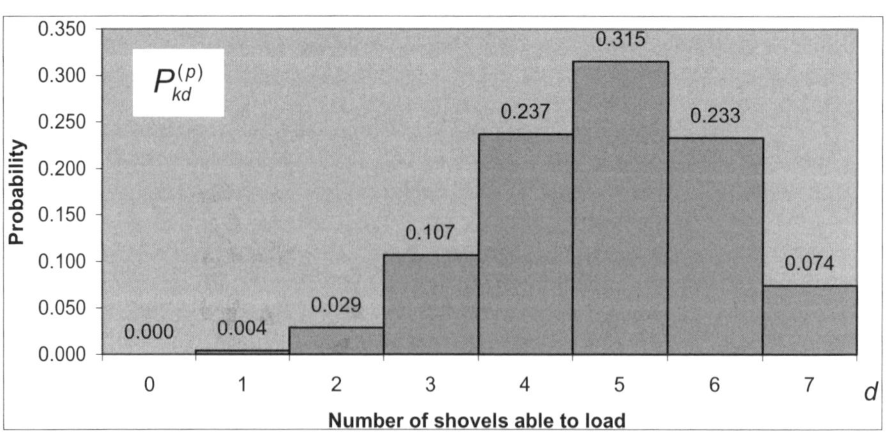

Figure 13.2. Probability distribution of number of shovels able to load for system $\mathfrak{S}_{XIV}$.

The conditional expected number of trucks for 7 shovels in work is:

$$E_{wk(7+0)} = \sum_{b}^{54} P_{wb}^{(p)} K_{566}(b)\left(\Psi_5(b)\int_0^7 xf_5(x;7,b)dx + \int_7^b xf_6(x;7,b)dx \right) = 5.24.$$

The ratio now needs to be considered:

$$(E_{wk(7+0)} / 7+0) = 0.75.$$

Looking at this figure, it looks as if the shovels are not heavily loaded from the point of view of serviceability. A recommendation can be formulated for the truck dispatcher:

move all possible trucks and direct them to the pit.

But this state is not so frequent—it only occurs in 7% of cases.

At the end of this stage the conditional expected value of time loss in the truck work cycle due to truck standstill waiting for loading needs to be calculated (sometimes this situation happens even when 7 shovels are able to load):

$$\Delta_{(7+0)} = Z' P_{kd=7}^{(p)} \left(1 - \sum_{g=0}^{7} p_{gd} \right) = 0.0 \text{ min}.$$

It can be seen that this event is a rare one.

- The next case that should be considered is when one shovel is down. The probability of this event is 0.233. Before the calculation is started, the mean loading time needs to be modified because one shovel failed and is replaced by a front-end loader. Applying formula (12.2), gives a new mean loading time

$$Z'_{n=1} = 2.8 \text{ min}.$$

The calculation is repeated, keeping the coefficient of variation $C_R = \text{const}$. This gives:

- The conditional probability that there will be no truck at 6 shovels able to load and 1 front-end loader:

$$p_{0(6+1)} = 0.000,$$

- Further conditional probabilities are:

$$p_{1(6+1)} = 0.000 \quad p_{2(6+1)} = 0.000 \quad p_{3(6+1)} = 0.002 \quad p_{4(6+1)} = 0.028$$
$$p_{5(6+1)} = 0.129 \quad p_{6(6+1)} = 0.249 \quad p_{7(6+1)} = 0.223 \quad p_{>7(6+1)} = 0.369.$$

- Conditional expected number of trucks at the loading system:

$$E_{wk\,(6+1)} = 7.27,$$

- Quotient

$$(E_{wk\,(6+1)} / 6+1) = 1.21$$

- Conditional time loss parameter:

$$\Delta_{6+1} = 0.2 \text{ min.}$$

Some comments are necessary here. The main difference compared to the previous point is that the mass of probability has been shifted a little away from zero. This displacement causes the conditional mean number of trucks at the loading system to increase and on average, both systems—loading and hauling—are well matched; the ratio is 1.21. Because of the stochastic nature of the exploitation process, some losses in time are inevitable—there is an increase in the conditional average time loss in truck work cycle due to time spent in the queue before the loading unit.

- In this third case, two shovels are down, i.e. there are two front-end loaders filling trucks. The probability of such an event is 0.315—i.e. the most frequent case. An appropriate adjustment to the mean loading time can be made, giving:

$$Z'_{n=2} = 3.5 \text{ min.}$$

The calculation is repeated:

- The conditional probability that there will be no truck at 6 shovels and 1 front-end loader able to load:

$$p_{0(5+2)} = 0.000$$

- Further conditional probabilities are:

$$p_{1(5+2)} = 0.000 \quad p_{2(5+2)} = 0.000 \quad p_{3(5+2)} = 0.000 \quad p_{4(5+2)} = 0.002$$
$$p_{5(5+2)} = 0.014 \quad p_{6(5+2)} = 0.049 \quad p_{7(5+2)} = 0.072 \quad p_{>7(5+2)} = 0.863$$

- Conditional expected number of trucks at the loading system:

$$E_{wk(5+2)} = 12.08$$

- Quotient

$$[E_{wk\,(5+2)} / (5+2)] = 1.73$$

- Conditional time loss parameter:

$$\Delta_{5+2} = 0.9 \text{ min.}$$

Looking at these results, some remarks are required. The displacement of the probability mass is visible, but not very great. There will be a longer queue of trucks at loading units, on average about 2 trucks. Such a situation is acceptable, although some trucks may be withdrawn from circulation.

- For the next case, three shovels are in failure, three wheel loaders work. The probability of this event was assessed as 0.237. The calculations are repeated and the results are as follows:
  - Mean loading time: $Z'_{n=3} = 4.1$ min
  - Conditional probabilities up to

$$p_{4(4+3)} = 0.000 \quad p_{5(4+3)} = 0.001 \quad p_{6(4+3)} = 0.004 \quad p_{7(4+3)} = 0.009 \quad p_{>7(4+3)} = 0.986$$

- Conditional expected number of trucks at the loading system:

$$E_{wk\,(4+3)} = 17.41$$

- Quotient

$$[E_{wk\,(4+3)} / (4+3)] = 2.49$$

- Conditional time loss parameter:

$$\Delta_{4+3} \cong 1 \text{ min.}$$

The exploitation situation in the pit changes gradually. The truck dispatcher may assess that an average of 17 trucks at loading units is too much; it is time to change the organization of the system and withdraw a certain number of haulers. The changes in the truck system can now be assessed. Looking at formula (8.2) compare the two expressions:

$$1+\frac{T_j}{Z'}=9.4 \quad \text{and} \quad 1+\frac{T_j}{Z'_{n=3}}=5.5.$$

This means that for this system in this situation, $(5.5/9.4) \times 100 = 58\%$ of the number of trucks are needed, ignoring their reliability problem. This result can be supported by analyzing the plots shown in Figure 12.1. But the calculation can be made simpler. There are an average of 17 trucks at 7 loading units. Thus, 10 units are not needed. If it is assumed that the truck dispatcher withdraws such a number of haulers from the pit,[1] then a system of $< 44, 32 >$ can be calculated. At first, the probability distribution of the number of trucks in work state will change. Applying the Maryanovitch model, gives:

$$P^{(p)}_{wj=44} = 0.959 \quad P^{(p)}_{wj=43} = 0.015 \quad P^{(p)}_{wj=42} = 0.010 \quad P^{(p)}_{wj=41} = 0.007$$

$$P^{(p)}_{wj=40} = 0.004 \quad P^{(p)}_{wj=39} = 0.003 \quad P^{(p)}_{wj=38} = 0.001 \quad P^{(p)}_{wj=37} = 0.001$$

and the average number of trucks in work state $E(44,32) \cong 43.9$.

Repeating the calculations gives:

- Conditional probabilities up to

$$p'_{4(4+3)} = 0.000,$$

$$p'_{5(4+3)} = 0.001 \qquad p'_{6(4+3)} = 0.006 \qquad p'_{7(4+3)} = 0.018 \qquad p'_{>7(4+3)} = 0.975,$$

- Conditional expected number of trucks at the loading system:

$$E'_{wk(4+3)} = 12.24$$

- Quotient

$$[E'_{wk(4+3)} / (4+3)] = 1.75$$

- The conditional time loss parameter:

$$\Delta'_{4+3} \cong 0.9 \text{ min.}$$

[1] Here the problem of how good this decision is is overlooked.

All the above parameters are marked by an apostrophe to indicate that these parameters are obtained for a system of changed organization.

This situation is similar to the second one. The truck dispatcher may withdraw some more units depending on the actual situation in the pit.

- The next case is: 3 shovels able to load, 4 shovels are in failure and 4 front-end loaders fill trucks. The probability of such an event is 0.107.

It should be noted here that this exploitation situation, or, to use another word, this state, has its own predecessor—the state when 4 shovels have been able to load supported by 3 front-end loaders in loading action. For this state the truck dispatcher decided to reduce the number of trucks accomplishing the transportation task to the system < 44, 32 >. Thus, the reasoning starts by assuming such a system.

The calculations give the following outcomes:

- Mean loading time:

$$Z_{n=4} = 4.7 \text{ min}$$

- Conditional probabilities up to $p'_{6(3+4)} = 0.000$ $p'_{7(3+4)} \cong 0.001$ $p'_{>7(3+4)} \cong 0.999$
- Conditional expected number of trucks at the loading system:

$$E'_{wk(3+4)} = 16.43$$

- Quotient

$$[E'_{wk(3+4)} / (3+4)] = 2.35$$

- Conditional time loss parameter:

$$\Delta'_{3+4} \cong 0.5 \text{ min.}$$

The situation is quite similar to the preceding one. Therefore it can be assumed that the truck dispatcher will withdraw, say, 10 trucks. This is the second change of organization of the system. Using the pair < 34, 42 >, and applying the Maryanovitch model, gives:

$$P^{(p)}_{wj=34} \cong 1$$

and the expected number of trucks in work state is obviously $E(34,42) \cong 34$.

If so, the further results of computations are:

- Conditional probabilities up to $p''_{4(3+4)} = 0.004$,

$$p''_{5(3+4)} = 0.047 \qquad p''_{6(3+4)} = 0.180 \qquad p''_{7(3+4)} = 0.264 \qquad p''_{>7(3+4)} = 0.505$$

- Conditional expected number of trucks at the loading system: $E''_{wk(3+4)} = 7.7$
- Quotient

$$[E''_{wk(3+4)} / (4+3)] = 1.1$$

- Conditional time loss parameter:

$$\Delta''_{4+3} \cong 0.3 \text{ min.}$$

- The next case is a situation where 5 shovels are down. The probability of this event is 0.029, and this case is assessed as rare. The state of the system as before is: < 34, 42 >.

Starting from the well-known set of parameters:

$Z_{n=5} = 5.3$ min, $p''_{5(2+5)} = 0.004\ p''_{6(2+5)} = 0.029\ p''_{7(2+5)} = 0.079$

$p''_{>7(2+5)} = 0.888$, $E''_{wk(2+5)} = 9.9$,

$[E''_{wk(2+5)} / (2+5)] = 1.4$, $\Delta''_{2+5} \cong 0.1$ min.

It appears that there is no special need to change the organization of the trucks.

- The penultimate case is a situation where 6 shovels are in failure. The probability of this event is 0.004, very rare. The starting point is an unchanged truck system.

  Parameters are as follow:

  $Z_{n=6} = 6$ min, $p''_{6(1+6)} = 0.001$ $p''_{7(1+6)} = 0.007$

  $p''_{>7(1+6)} = 0.992$, $E''_{wk(1+6)} = 12.32$,

  $[E''_{wk(1+6)} / (1+6)] = 1.76$, $\Delta''_{1+6} \cong 0.2$ min.

  The truck dispatcher can make the decision to withdraw some trucks to reserve. However, this state should not last long, so an unchanged system is assumed.

- The last case is a state when all shovels are unable to load; only front-end loaders execute their duties. The probability of this event is below 1%.

  The parameters are as follows:

  $Z_{n=7} = 6.6$ min, $p''_{>7(0+7)} = 1$, $E''_{wk\,(0+7)} = 14.38$,

  $[E''_{wk\,(0+7)} / (0+7)] = 2.05$, $\Delta''_{0+7} \cong 0.2$ min.

  It is assumed that the truck dispatcher does not interfere.

Now is time to release the considerations from the conditional approach.

Now the unconditional measures can be calculated.

The number of trucks that can be loaded simultaneously by the shovel system equals n maximum. Therefore, the unconditional probabilities can be computed from the formula:[2]

$$p_d = \sum_{i=0}^{n} P_{ki}^{(zd)} p_{d[i+(n-i)]}. \tag{13.2}$$

The results of the calculation for the example analyzed are as follows:

$p_0 = 0.000$ $p_1 = 0.000$ $p_2 = 0.000$ $p_3 = 0.003$
$p_4 = 0.022$ $p_5 = 0.068$ $p_6 = 0.115$ $p_7 = 0.792.$

This probability distribution is shown in Figure 13.3.

The expected number of trucks loaded by the system of loading machines, i.e. shovels and front-end loaders is:

$$E_{wlk} = \sum_{d=1}^{n} dp_d = 6.7 \quad \text{trucks.}$$

This measure takes into account the reliability of the shovels, the reliability of the trucks, and the variable organization of the whole machinery system.

[2] Notice the difference between probability patterns (10.10)–(10.12) and (12.4).

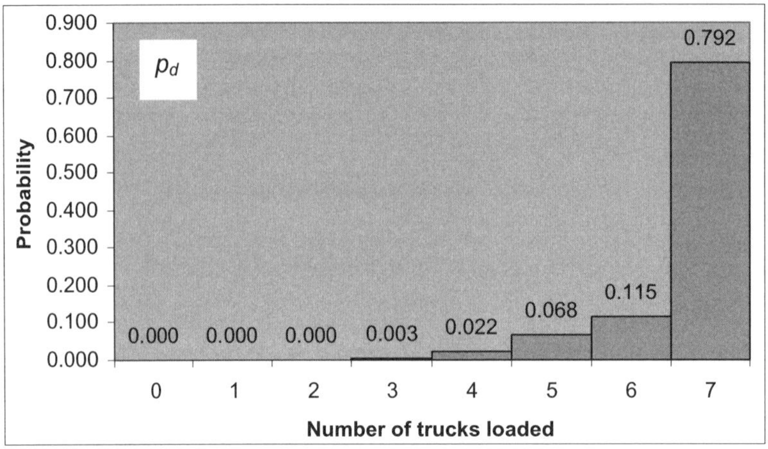

Figure 13.3. Probability distribution of number of trucks loaded by system $\mathfrak{S}_{XIV}$.

Looking at Figure 12.5 a short comment should be made. In almost 80% of cases all 7 loading points will work. However, in more than 10% of cases 6 loading machines will fill trucks. In almost 7% cases there will be only 5 loading points, although a constant number of loading machines—7—is assumed. The reason for this phenomenon is the occasional lack of hauling machines at loading units, though the average truck queue length is quite high, as will be proved by the next step of the analysis.

Observe that, coming from the definition of expected value (p. 71), the mean number of shovels able to load—that is 4.8 units—can be calculated, and therefore an average of 2.2 front-end loaders will be executing their duties in the system.

The unconditional expected number of trucks at the system of machines able to load is:

$$E_{wlk} = \sum_{d=1}^{n} E_{wkd} P_{kd}^{(zd)} = 9.9 \qquad \text{trucks.}$$

Therefore, the average number of trucks in a queue per one loading machine:

$$\theta = 1.46 \qquad \text{trucks.}$$

Looking at this figure, it may be that in some cases the truck dispatcher should have to withdraw some trucks from circulation.

Applying again the cited definition of the expected value, the average number of trucks in work state in the system can be calculated—that is: 46.8 trucks.

The mean time of truck work cycle can be calculated from an equation formulated in symbolical form:

$$T_c = T_j + \sum_i Z_i P_{ki}^{(zd)} + \sum_i \Delta_i \tag{13.3}$$

giving: $T_c = 18.5 + 3.6 + 2.8 = 24.9$ min.

The above equation gives the following information. The average time of a truck work cycle consists of the mean time of truck travel (haul—dump—return), 18.5 min, plus the mean time of truck loading, 3.6 min, plus the mean time lost due to trucks waiting in a queue for loading, 2.8 min. Looking at the new mean loading time it can be concluded that the average value is longer by more than 60% compared to the mean loading time by shovel. This average of 2.2 front-end loaders has made this so.

Some output measures of the system can now be calculated.

The theoretical output of the shovel system:

$$W_{tk} = \frac{60}{Z'} n = 190.9 \qquad \text{trucks/h.}$$

Taking into account the shovel's reliability, gives the potential output of this system:

$$W_{pk} = \frac{60 A_k}{Z'} n = 160.4 \qquad \text{trucks/h.}$$

If the shovel's accessibility is considered, its own output can be obtained:

$$W_{ok} = \frac{60 G_k}{Z'} n = 131.5 \qquad \text{trucks/h.}$$

Now the wheel loaders can be included and the whole loading system effective output can by calculated, given by the formula:

$$W_{efk} = \frac{60n}{\sum_i Z_i P_{ki}^{(zd)}} = 116.7 \qquad \text{trucks/h.}$$

The truck system output, ignoring time losses, is

$$W_{pw} = \frac{60}{T_j + \sum_i Z_i P_{ki}^{(zd)}} E_w = 127.1 \qquad \text{truck work cycles/h.}$$

Taking into account truck queues and time losses, gives

$$W_{efw} = \frac{60}{T_c} E_w = 112.8 \qquad \text{truck work cycles/h.}$$

It appears that in further analysis, the system effective output can be assumed to be approx. 112 trucks/h.

Finally, one interesting system parameter can be calculated, that is a difference between two expected values:

$$E_w - E_{wk} \cong 37 \qquad \text{trucks.}$$

On average, such a number of trucks will be just in the travel state. Knowing which part of the travel state is associated with movement outside of the pit, the average number of haulers that will circulate in the pit can be evaluated.

Looking at the above results of calculation, several interesting conclusions can be drawn.

If the output W_{tk} of the shovel system only is compared with the whole system's effective output, and taking into consideration the shovels' reliability and accessibility; their capability of loading was assessed at 131.5 trucks/h, ignoring the employment of spare loaders. The system effective output in which spare loaders have been included was assessed at 112 trucks/h. This means that approximately 15% of this capability has been lost. This percentage vanished mainly for two reasons. In the capability of loading mentioned, the reliability of trucks was not taken into account; the reliability that was assessed was low. Moreover, changes in the organization of the system were not taken into consideration. These two factors generate truck queues at loading machines and subsequently losses.

Generally, an interesting case would be a comparison between two systems: one employing shovels only without a spare loading unit, and another using front-end loaders to replace failing shovels. There is no doubt that changes in productivity would be visible, but the difference between these two outputs depends strongly on the proportional coefficient τ.

It should be realized that the whole analysis and calculations are valid for a given phase of mine development. When a pit enlarges hauler routes will elongate and more machines will be needed. For a different number of machines and different values of input parameters the whole calculation procedure must be repeated from the beginning.

CHAPTER 14

Systems with priority and the ideal dispatcher

14.1. INTRODUCTION

Earlier it was stated that during exploitation of a mineral deposit it sometimes becomes necessary to speed up removal of waste from the pit or to accelerate haulage of ore. This means that the truck dispatcher must first direct empty hauling units coming to the pit, to loading machines serving an appropriate load. If it is assumed, for unambiguous and clear further considerations, that waste must be removed. The loading system is now divided deterministically into two subsystems, one of which has *priority*. The hauling system is also divided—now stochastically, into two subsystems.

The fact that priority is an issue disturbs the regular exploitation process of the machinery system applied. System characteristics change. There is currently a lack of publications describing analytically the changes that can be expected if this priority is implemented in the current control of the system. A system where such a type of priority is introduced is now considered.

Today, the control of every larger machinery system of this kind in the world is assisted by computer systems. Truck routes are continuously monitored and each machine possesses the ability to communicate directly with the dispatching centre. There is also communication with the back-up facility and parking area. Frequently TV cameras are located in the most important points, such as the unloading area, points where trucks equipped with a trolley assist system join the overhead wire, etc. There are monitors in the dispatching centre showing what is going on at these points and monitors showing either numerically or as plots the parameters and characteristics of the machinery system. A central computer continuously gathers system data. Historical information is stored on the discs. In the centre there is almost all possible information on the working area, the three-dimensional geometry of the pit, current positions of all machines, their status, what is happening in the dumping area and the unloading station of ore. Often computers have professional graphical animation programs to simulate past, current or future processes running in the mine. Some programs support decision-making in situations such as a blockade of part of the haul road by blasted rock or slope failure, the appearance of heavy rain, etc. New programs in this field are being tested at present.

Every decision of the truck dispatcher—changing the number of circulating machines, the speed of haulers in the pit, positions of unloading points, etc.—changes the machinery system characteristics. If this change is short lasting or not great, its influence on the system characteristics can be ignored. However, if this change is significant, long lasting or permanent, it is worth trying to evaluate the significant changes that can be expected in the values of the basic system parameters. Mine centres estimate these changes based on the principle: 'income—outcome'. They keep historical data in the computer and take into account the numerical characteristics of a given state of the system. For each different state of the system, there was a different number. No analytical mechanism is being traced, and there is no explanation as to the parameter value for a given state. Forecasting the situation is assessed on the simple projection principle. Outcomes obtained using this method of reasoning are sometimes good and sometimes far from reality. Very often, the rich experience of the decision-maker is used to adjust these outcomes.

Returning to the main point of the considerations, it is time to find an answer to the following question: what is going to happen in the machinery system exploitation process when the truck dispatcher starts to systematically direct incoming trucks to the loading machines connected with overburden removal?

Before this question can be answered, it should be realized that the considerations so far on the exploitation of the machinery system have been organized in such a way that the service load of all power shovels is approximately the same. In other words, it has been assumed that, on average, all trucks will visit all power shovels approximately the same number of times in a long time. This, while obviously a simplification in comparison to mine practice, can be accepted. It also means that if there are m trucks in the system and n power shovels it can be assumed that (m/n) trucks serve each shovel, on average in the end. The quotient

$$\zeta = m/n$$

will be called the *service rate*. This also means that if in the system the same number of power shovels load waste and the same number of shovels load mineral, the output obtained from the system will be approximately half and half.

After the dispatcher's decision based on priority, the loading system is divided into two subsystems: a set of machines $\mathfrak{h}_m$ of power n_m loading mineral and a set of machines $\mathfrak{h}_w$ of power n_w loading waste; $n_m + n_w = n$. This division is deterministic in nature.

The second possible change after implementation of priority could be in the mean travel time of the truck. In some cases, the average time of truck travel from the mineral loading machine may be different to the mean time of truck travel from the machine loading waste. Usually this difference is not significant and can be ignored. However, if the empirical data shows that the difference between these means is considerable; this fact must be included in the calculation procedure. Doing so does not cause any problem.

The main difference will be connected with the service rate value. Generally, this proportion is specific for a given machinery system at a given stage of mine development. If the priority principle is implemented and the machinery system operates, the value of service rate changes. It is known that to improve the intensity of waste removal the dispatcher will ensure that the power shovels loading mineral are under loaded from the service point of view, and power shovels loading waste will be of a high value of service load. To write this fact conventionally:

$$\mathfrak{H} \cup \mathfrak{D} : \mathfrak{h} = \mathfrak{h}_n \cup \mathfrak{h}_m : \zeta_w \uparrow, \zeta_w \downarrow ; n_n + n_m = n$$

where D denotes the decision on the implementation of the priority.

The problem now is how great these changes will be.

These changes depend on two factors:

- Quality of the machinery design
- Quality of the machinery control, mainly truck dispatching.

If the system is properly selected, i.e.

- The number of trucks directed to accomplish the transportation task and the number of trucks in reserve suitably matched to the exploitation requirements determined by the productivity of power shovels, and
- The number of repair stands in the shop is such that the queue of failed trucks waiting for repair is negligible

the operation utilization of machines of the system should be high. This means the possibility of increasing significantly the productivity of the system is not very great. Moreover, if the output of some shovels is raised, then inevitable losses in some areas of the exploitation process will simultaneously be observed. It concerns, first, longer truck queues before shovels. Here it needs to be considered whether these small changes will be enough to match the mine requirements in terms of productivity. It depends on the assumed production plans of the mine. Sometimes it can be enough, but sometimes not. If not, there are two basic solutions.

The first is to direct one additional loading machine for waste removal.

The second solution in this regard is to purchase a few more trucks, thus going, incidentally, against the future needs of the mine. If the machinery system is properly selected this type of operation makes the values of some of the parameters of system performance worse. An increment will be observed in both the productivity of the shovel system and the average truck queue length.

If the machinery system is not properly selected—the proof of this fact is visible in both truck wasting time in queues for loading and shovel wasting time waiting for hauling machines—the introduction of the priority in truck dispatching will make significant changes in the exploitation process of the system. A drastic increment in the productivity of the shovel subsystem loading waste will be noticed, as well as rapid decrement in the output of shovel subsystem loading ore.

Changes in the service rate value are a function of the quality of control of the system. It is obvious that the more careful and correct the truck dispatching following fast changes in the exploitation situation in the pit, the higher the efficiency of the system control. It is assumed here that decisions made by the truck dispatcher are rational ones.

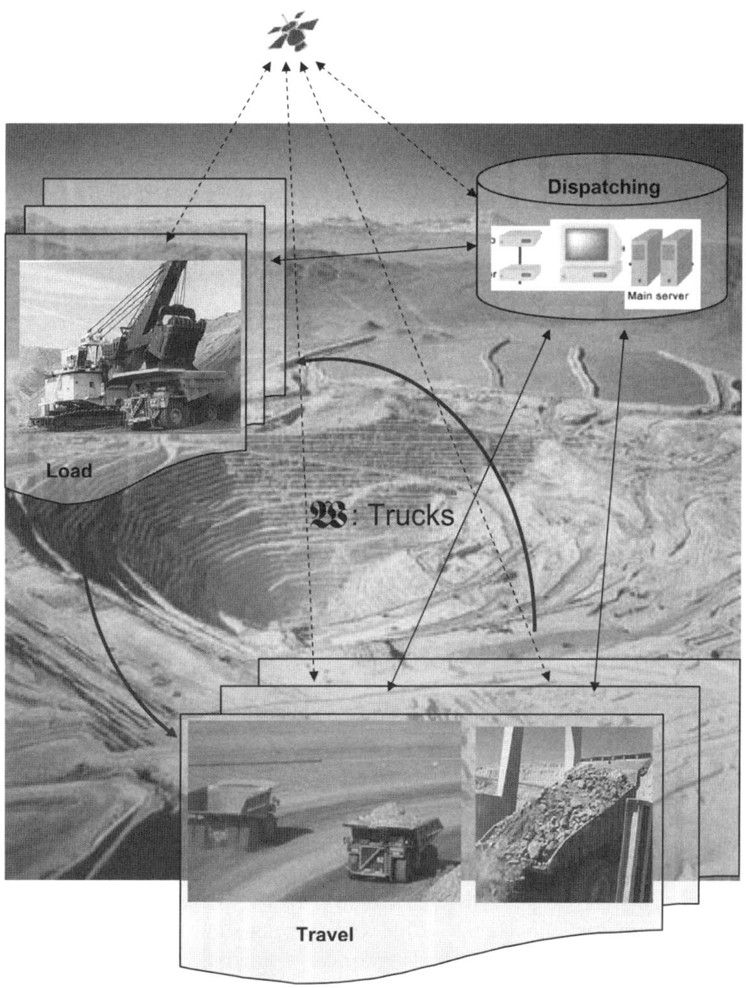

Figure 14.1. Operating scheme of a shovel-truck system and dispatching centre with the assistance of a GPS system.

To summarize:
The real changes in the rate service value are only possible to estimate for a given system, for a given stage of mine development and for a given truck dispatching practice. The last component depends on the individual properties of a dispatcher.

After this statement, the question can be asked, whether it is possible to estimate these interesting changes in spite of the fact that they depend on the subjective properties of the dispatcher. Fortunately, the answer is yes.

14.2. MODIFICATION OF CASE II—THE IDEAL DISPATCHER

The procedure presented in the previous section is now considered.

In chapter 13, the case of the study was orientated towards changes in the calculation procedure due to the application of spare loaders. Now how the procedure should be modified due to the implementation of priority, or in other words, how the effects of the truck dispatcher's decisions can be traced, needs to be considered.

The system $\mathfrak{S}_{XIV}$ and its properties are known. First, the number of power shovels that load waste needs to be specified. This is important for the further considerations. It is assumed that, e.g. 4 power shovels serve overburden removal and the remaining 3 machines load mineral.

The number of trucks directed to accomplish the transportation task has been fixed and the reserve size determined. The number of repair stands has also been evaluated.

The procedure starts by discussing the case when all power shovels are in the state of possibility of loading. The following data repeats some results.

The conditional probabilities and their outcome were computed as follows:

$$p_{1(7+0)} = 0.000 \qquad p_{2(7+0)} = 0.002 \qquad p_{3(7+0)} = 0.035 \qquad p_{4(7+0)} = 0.200$$
$$p_{5(7+0)} = 0.378 \qquad p_{6(7+0)} = 0.276 \qquad p_{7(7+0)} = 0.087 \qquad p_{>7(7+0)} = 0.021.$$

Remember that $p_{g(d+s)}$ is the conditional probability of an event that will be g trucks at d shovels able to load and s front-end loaders able to load. Obviously, $d + s = n$.

The first significant probability here is $p_{2(7+0)}$. If the dispatcher acted without any mistake these 2 trucks were directed to power shovels loading waste material. This means that no mineral is being loaded.

The next case of significant probability is the situation where 3 trucks are at loading machines. Again, the dispatcher made sure that these machines are at the proper loading shovels. The result is 3 trucks are filled by waste and no trucks filled by ore.

The third case to consider is 4 trucks are filled by waste and no truck filled by mineral. The probability of this case is $p_{4(7+0)} = 0.200$.

In the next case there is one truck loaded by mineral if the dispatcher made the right decision.

In further cases, the number of trucks connected with waste remains unchanged, and the number of trucks loaded by ore increases up to the case when 4 trucks are connected with waste and 3 trucks with mineral. The corresponding probability is $p_{7(7+0)} = 0.087$. The last case has an identical division of trucks.

After these considerations, the following important conclusions can be drawn.

First, the decisions made by the truck dispatcher can be expressed by appropriate exploitation system parameters.

Secondly, several times it has been assumed that the dispatcher's decisions were made without any mistake. This in turn assumes that the dispatcher is an *error-free* person, i.e. an ideal one. The **ideal dispatcher**—in the case considered here—is one who does not allow a situation to occur whereby a truck is loaded with non-priority material and a priority-loading machine is waiting for a truck.

To make the dispatcher more real, an assessment of his mistakes should be introduced; in some cases the truck should go to a different destination where it should not go. This introduction depends entirely on the person conducting the modelling.

These considerations can now be taken further.

To get an idea what kind of changes has been made by the truck dispatcher, the expected values for the number of loaded trucks need to be compared.

a. If there is no priority the mean number of loaded trucks is: 5.21. Knowing that 4 trucks are involved in waste loading and 3 trucks in mineral loading, the mean number of loaded trucks by each type of material can be calculated. The outcome is:

   $(4/7) \times 5.21 = 2.98$ trucks of waste and $(3/7) \times 5.21 = 2.23$ trucks of mineral.

b. If there is priority the mean number of loaded trucks is:

   3.96 trucks of waste and 1.25 trucks with mineral.

Thus, the effect of the introduction of priority in the removal of waste can be estimated. In the case when all 7 shovels are able to load (no spare loading machine engaged) instead of loading approximately 3 trucks with waste, almost 4 trucks will be loaded with this type of material. This one truck is taken from loading mineral.

Consider the ratio: the mean number of loaded trucks per one loading machine.

For the system of power shovels loading waste:

$$3.96/4 = 0.99$$

for the system of power shovels loading mineral:

$$1.25/3 = 0.42.$$

The same ratio calculated in the previous chapter (no priority) is 0.75. The changes made by the truck dispatcher are clearly visible.

In a similar way, other values can be calculated by considering the next cases of the calculation procedure.

It is worth realizing that the truck dispatcher changes the allocation of trucks, the service load of some machines, productivity of subsystems of loading machines, but does not change the properties of the machines. Nevertheless, if an implemented rule, like the one concerning priority, is long lasting and the intensity of usage of some machines is significantly different, it can influence the intensity of failures of some machines. The reliability of these machines will change. However, as for power shovels, they are robust machines, usually of relatively high reliability; they can work with higher intensity for a longer time and changes in the intensity of their failures will not be observed. This is the case for machines made by good producers.

CHAPTER 15

Hauling distance and system characteristics

At least two vital problems connected with shovel-truck systems are still waiting for comprehensive, penetrative analysis, namely:

- Determination of a formal principle describing when enlargement of the truck fleet should be made as the function of development of the mine
- Construction of a mathematical model of truck dispatching rules that will allow system calculation.

Both problems are vital from a practical and a theoretical point of view. However, their difficulties are significantly different. In spite of the fact that there have been several dozen interesting publications concerning truck dispatching, there has yet to be a paper presenting such a description of truck dispatching that can be used to trace changes in the exploitation process of the system and calculate essential efficiency parameters of the system at any moment of time. Such a model will permit comparison between particular rules and an assessment of which rule is better or worse from a given point of view.

The second problem relies on the construction of a formal principle indicating when the enlargement of a truck fleet should be made due to an enlargement of the pit. This enlargement means an increase in haulage and return distances. This problem is complicated by at least two factors. The first is connected with the pure technical and exploitation relationships. The second is associated with the financial policy in operation at the mine. Therefore, the first reason can be discussed in the scope of this monograph; the latter will not.

In these considerations, therefore, the relationships between the increasing mean time of truck travel and basic system parameters will be discussed. The plots presented will show some tendencies in changes and graphical relationships between parameters. Based on formulas given in this book the Reader may calculate his or her own case.

Recalling formula (8.2).

$$h = nG_k \frac{Z' + O + W + R}{Z'}.$$

When the mine develops and the hauling distance increases, two variables in this formula will change—that is O and R—as well, obviously, as h. Thus this expression can be rewritten in the following form:

$$h = nG_k \left(1 + \frac{T_j}{Z'}\right) = \frac{nG_k}{Z'} T_c$$

where $T_j = O + W + R$ is the mean time of truck travel.

Because the magnitude h indicates the number of trucks in work state, this relationship is modified by introducing the truck steady-state availability A_w to give:

$$V = \frac{h}{A_w} = \frac{E(\boldsymbol{D})}{A_w}.$$

The number of trucks needed is a natural number, thus the following is calculated:

$$\nabla(V) = \nabla\left(\frac{nG_k}{Z'A_w}T_c\right) = \nabla\left(\frac{nG_k}{Z'A_w}\left(1+\frac{T_j}{Z'}\right)\right), \quad \nabla(V) \in \mathfrak{N}, \tag{15.1}$$

and this function is stepped in character like the one shown in Figure 15.1.

The above expression is the equation of a stepped function supported on a straight line. The proportional coefficient

$$\frac{nG_k}{Z'A_w} = \tan\varsigma \tag{15.2}$$

is directly proportional to the number of shovels in the system and their loading capability, and indirectly proportional to the mean loading time and truck steady-state availability.

Looking at formula (15.1), it easy to notice that if there is only one shovel and it is totally reliable, by continuously loading (that means $G_k = 1$), an additional unit should always be added to the system when the increment in the mean truck travel time exceeds the mean loading time.

Analyzing formula (15.1), it can be stated that:

- The increase in the number of trucks needed is directly proportional to the number of loading shovels as well as to the shovel ability for loading
- The increase in these magnitudes n, G_k makes the increase in the angle ζ and shifts the whole function up
- The increase in the mean loading time and/or increase in the steady-state availability of truck cause a decrease in angle ζ, so the number of trucks required decreases.

If this is displayed in a graphical form, Figure 15.2 is obtained.

Further regularities can be traced by looking at the results of the calculation of example machinery systems.

Now the following shovel-truck system is considered:

$$\mathfrak{S}_{\mathrm{XIII}} :< \mathfrak{H} : n = 5, A_k = 0.88, B_k = 0.80;$$

$$\mathfrak{W} : A_w = 0.665, 0.706, 0.794\ 0.828;$$

$$\delta = 0.077, 0.068, 0.044, 0.0406,\ \sigma_p = \delta^{-1},\ \sigma_n = 0.35\ \gamma^{-1}; \mathfrak{Z} : k >$$

with the mean adjusted loading time: $Z' = 2.5$ min.

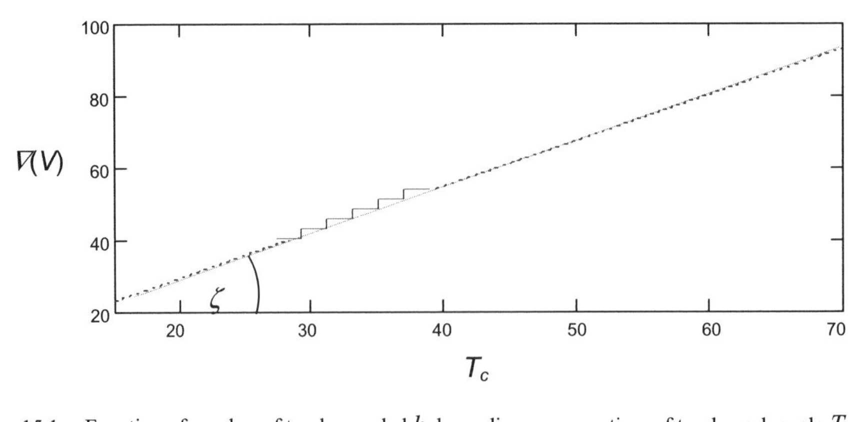

Figure 15.1. Function of number of trucks needed h depending on mean time of truck work cycle T_c.

This system verbaly can be defined verbaly as follows. System $\mathfrak{S}_{XIII}$ consists of 5 loading shovels of steady-state availability 0.88 and accessibility 0.80. Selected trucks are of four different reliability levels, starting from low reliable ones of 0.665 steady-state availability, through 0.706 and 0.794 up to the most reliable of 0.828 steady-state availability. The intensity of truck failures δ is assessed as: 0.077, 0.068, 0.044, 0.0406 h^{-1}, respectively. The intensity of truck repair is the simple function of A and δ. The probability distribution of work times for trucks is exponential, thus the standard deviation equals the mean. The standard deviation of repair times is assumed stable for all trucks and equals 0.35 of the appropriate mean.

The following is a discussion of the results of calculations for the data system.

Figure 15.3 shows the number of trucks needed for the system as the function of the mean time of truck travel. Truck reserve size is not taken into account.

Looking at this figure one totally astonishing conclusion can be formulated:

The number of trucks needed for circulation almost does not depend on steady-state availability of the truck.

This conclusion looks to be against the elementary, verified principle of reliability theory stating that for more reliable machines, fewer of these machines are needed to accomplish the formulated task.

Immediately a crucial question comes into being: why has such regularity occurred?

The plot presented at Figure 15.3 was obtained from the calculation procedure shown at Figure 8.1. Formulas (8.2) and (8.5)/(15.1) were applied. From formula (8.5) the information on

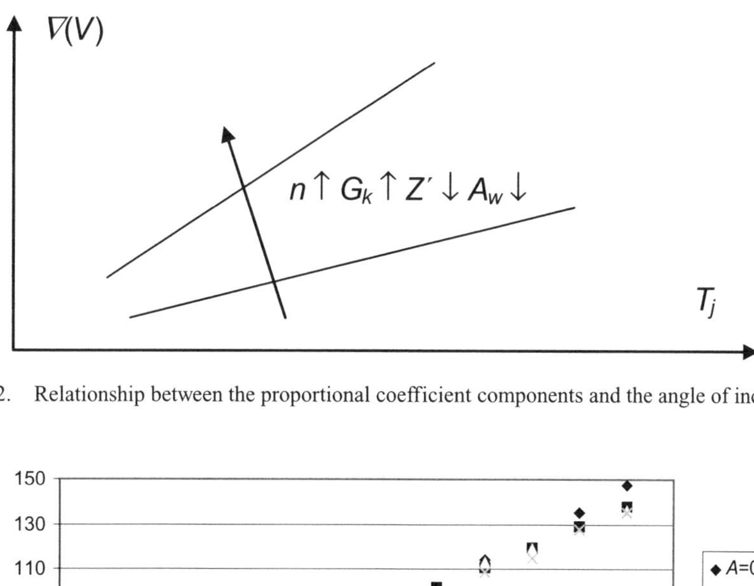

Figure 15.2. Relationship between the proportional coefficient components and the angle of inclination ζ.

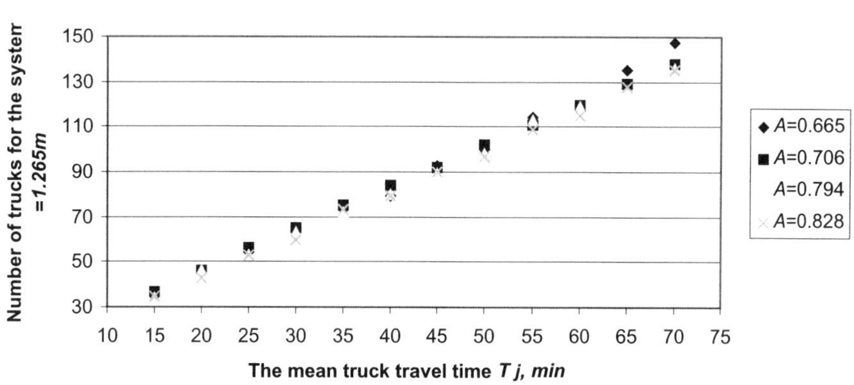

Figure 15.3. Function of number of trucks needed for the system depending on the mean time of truck travel T_j.

the minimum number of trucks in work state is obtained whereas the next formula gives the total number of trucks that should be employed. If so, now the Maryanovitch model can be applied to get the division of trucks that should be directed to accomplish the transportation task formulated by the shovel subsystem (this is included in these formulas)—*m* and trucks in reserve—*r*. This division is according to the conditions given in formula (8.6); recall: the total number of trucks applied should be the minimum, the reserve size should be the maximum. And just this scheme ensures that the effect of the variable availability of trucks applied is visible in the reserve size almost exclusively.

Additionally, one very important inference for mine practice can be formed:

The applied method for the selection of the number of trucks in the system with division on the number of vehicles directed to the pit and directed to the reserve, ensures the minimum number of trucks that will be directed to the pit. This is very convenient due to many reasons that were listed on page 77.

The statement that truck availability has an influence on the reserve size needs to be verified.

The results of calculations in this regard are presented in Figure 15.4.

A few conclusions can be drawn by looking at Figure 15.4:

- The truck reserve *r* strongly depends on the availability of trucks
- The truck reserve *r* should increase when the steady-state availability of truck A_w decreases
- The truck reserve *r* is directly proportional to the mean time of truck travel T_j
- Points obtained from calculations create approximately straight lines, but the dispersion of these points is in some cases high
- The angle of inclination of these lines becomes steeper for lower reliability of trucks.

The relationship between the number of repair stands needed for a given truck system versus the mean time of truck travel is presented in Figure 15.5.

Analyzing information contained in these plots again calls to mind three remarks:

- The relationship between the number of repair stands needed for a given truck fleet determined by the mean time of truck travel is clearly linear for all levels of availability.
- The straight line connected with lower reliability lies above the straight line of higher reliability.
- The slope angle for the line increases for decreasing reliability of machinery.

Figure 15.6 illustrates the ratio of the number of trucks in the system (1.265 *m*) to the number of trucks in reserve versus the mean time of truck travel T_j. Observe that this ratio is the quotient: number of trucks in operation per 1 truck in reserve.

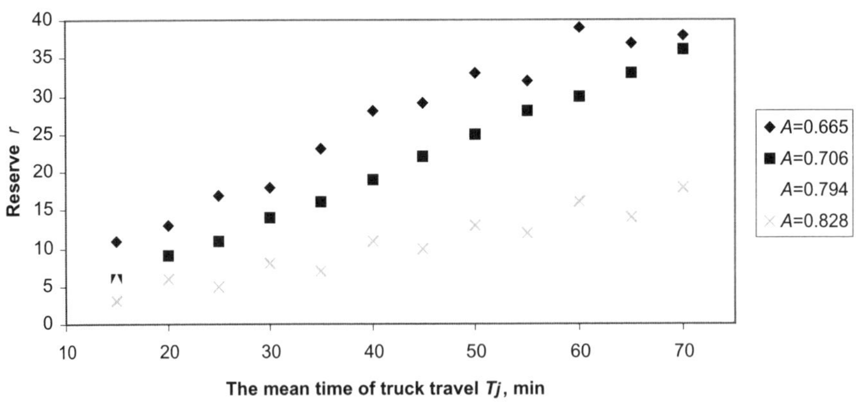

Figure 15.4. Function of truck reserve *r* depending on the mean time of truck travel T_j.

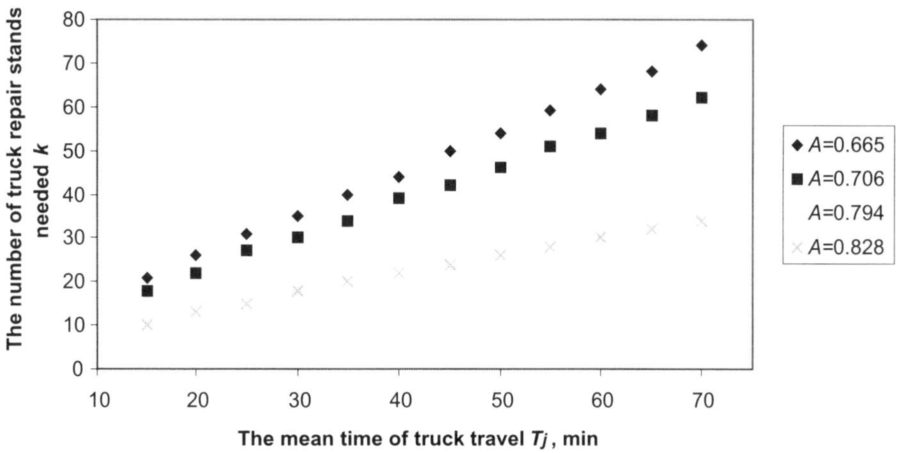

Figure 15.5. Function of number of truck repair stands needed for the system k depending on the mean time of truck travel T_j.

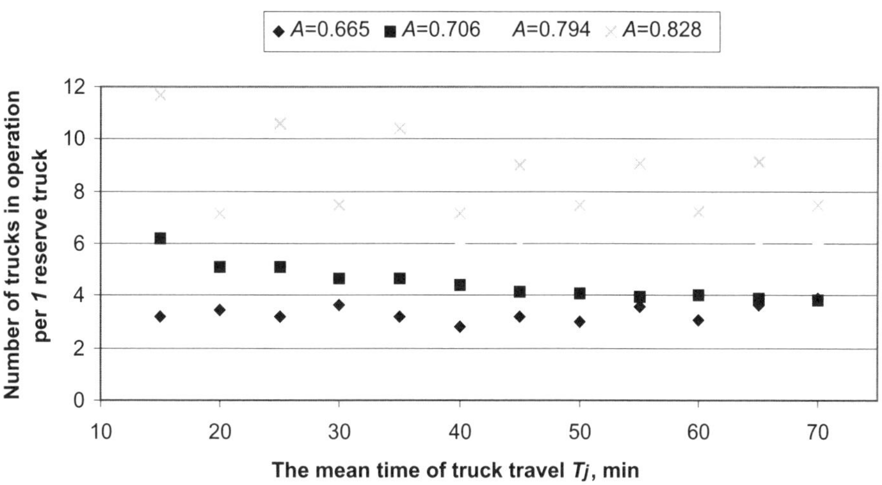

Figure 15.6. Function of number of trucks in operation per 1 truck in reserve depending on the mean time of truck travel T_j.

The information contained in the above graph is interesting. Some significant conclusions can be drawn, namely:

- For trucks of low reliability the situation is rather steady but quite unfavourable:
  - 1 spare unit is needed for every 3 to 4 trucks in operation for $A = 0.665$
  - for $A = 0.706$ it is recommended to have 1 spare for every 4 to 6 units depending weakly on the mean time of truck travel, less spares for longer trips (that is a larger system of trucks)

For trucks of good or very good reliability it is recommended to consider each case rather carefully and:

- for high reliability $A = 0.794$, approximately 1 spare is needed for every 6 to 9 units
- for very high reliability $A = 0.828$, 1 spare is needed for every 7 to 11 units;

Therefore Hartman's recommendation (1987, p. 256) appears to be lacking

- When the mean time of truck travel increases, i.e. the number of trucks employed in the system also increases, the ratio of the number of trucks in operation to number of trucks in reserve slightly decreases for $A > 0.700$; for $A = 0.665$ there are no grounds to reject the statistical hypothesis stating that the sequence of ratio values is constant.

Analyzing Figure 15.4 and especially Figure 15.6 the question can be formulated: for what reasons are the points so dispersed? There are at least two grounds for this dispersion. First, an accumulation effect is visible, made by rounding up some values, as too is an effect of searching for optimal values in the discrete space (Maryanovitch model). The next reason for the points spreading on Figure 15.6 is the variable load of the repair shop (the heavy traffic condition). In addition, it is necessary to remember that the calculation procedure gives approximate results only, although it does seems that the results are good enough to be applied in mine practice.

CHAPTER 16

Special topic: Availability of a technical object

At all stages of analysis, modelling and calculation a very important reliability parameter is needed, which is called availability. In connection with this term and the history of its evolution in English-language mining engineering literature, some explanations are indispensable.

The starting point is the definitions that can be found in different papers in the mining area.

As early as 1969, Connell (see also 1973 pp. 18–22 and 18–23 in *SME Mining Engineering Handbook*) stated: 'Difficulty can arise where various definitions of truck availability are used. The availability factor … requires the following definition: the difference (possible hr—downtime hr) divided by possible hr. Other terms are in use, such as mechanical availability, tire availability, etc. Such factors can be calculated as follows: mechanical availability is the difference (possible hr—mechanical downtime hr) divided by possible hr.'

Comments:

It is difficult to discuss the above definitions because no formula has been given to determine *the possible hr*. Besides, the term *available hr* is also given along with no relationship between these two terms.

'Availability Factor—A measure of the reliability of machines as regards freedom from mechanical failures. It is the actual working time divided by the available working time, in hours. An availability factor of 95% is excellent, 90% is good, and 85% is acceptable' (Church 1981, p. G-2).

Comments:

Why are mechanical failures so important that the whole availability measure is connected with mechanical collapse exclusively?

If the time of observation of the object operation is relatively short this ratio gives poor assessment.

It is necessary to consider different estimators for components of the ratio; these components are estimates of random variables, and the pattern of a particular estimator depends on the plan of investigation applied. Therefore in some cases the above estimate is good, and in others quite poor (see for example Gnedenko *et al.* 1969, Czaplicki 1990a). Comments associated with the mathematical statistics given here remain valid for further citations and will not be repeated.

This approach neglects entirely the intensity of machine work; it makes no sense to compare two estimates of this parameter for two of the same machines if one has been heavily used and the second one has not. For machines like the ones considered in the paper, the machine is in a work state waiting in a queue for loading. The longer the time spent in a queue, the lower the intensity of usage of the machine.

'Mechanical availability is the availability after mechanical repair, preventive maintenance, and servicing have been accounted for' (Dohm 1992, p. 1284, SME MEH).

Comment:

From a mathematical point of view, this statement means almost nothing.

'Mechanical availability is the ratio of the number of working hours of a machine to the sum of the number of working hours and the number of repair hours' (Korak and Müller 1987, p. 601).

Comments:

Why is this availability parameter called *mechanical*?

What about a machine consisting of many electrical or electronic subsystems?

Perhaps only the ability of a certain part of the machine is being assessed? Why?

'Physical availability is the ratio of the sum of the number of working hours of a machine plus standby hours to the total number of hours' (Korak and Müller 1987, p. 601).

Comments:
The measure defined here is an exploitation parameter measuring machine ability to fulfil requirements in its exploitation process.
In this paper the equivalent parameter for shovels is the accessibility coefficient (factor). For trucks this parameter is not defined because it is not needed in the procedure.

'Usable availability is the ratio of the number of working hours of a machine to the sum of the number of working hours and the number of standby hours' (Korak and Müller 1987, p. 601).

Comments:
This exploitation parameter gives information about what degree the machine is used for work taking into account its ability to work. However, the number of standby hours depends on the structure of the system in which the machine operates, the system control (here: method of truck dispatching) and some other factors. Therefore, comparison values of this parameter for two machines operating in two different systems make no sense. For two machines operating in the same system such a comparison will in some cases make sense, but not in others.

These authors produced many different types of availability, but a definition of what availability is *in general* is lacking.

'Availability: This parameter reflects the decrease in scheduled time due to mechanical and electrical delays. A delay for a mechanical or electrical problem is scheduled time that the machine cannot operate because of a mechanical or electrical failure or repair; this includes wear part replacement, welding, etc., as well as major repair work. It does not, however, include delay time for mismanaged transportation of parts or labour … the availability may be calculated by dividing the hours available by the hours scheduled.' (Humphrey 1990, p. 651).

Comments:
Why is there a division into failure and repair? The first is an event, the second is a state characterized by a random variable.
What does *hours available* mean? The definition was not given.

'**Availability and Utilization**. Availability and utilization do not have standardized definitions and, thus, these terms must be used with care.' [Further on in the text there is no definition of availability given and no measure of it] (Hays 1990, p. 684).

Comment:
Better, no comments.

'The conventional definition of mechanical availability of trucks is the ratio between working hours and the sum of working hours and repair hours. Working hours include the waiting times at shovels and the dump-crusher and the repair hours include the waiting time outside the repair shop. Since these waiting times are dependent on the system which is being studied, it seems appropriate to introduce a more basic availability measure, from which the conventional mechanical availability and the effective utilization can be worked out. Such a measure could be called the "internal availability" defined as the ratio between functioning hours and the sum of those hours and the effective repair hours. None of the hours should include waiting time at shovels, dump-crusher, or outside the repair shop.' (Elbrond, J., 1990, p 746).

Comments:
Again the question: why is the term *mechanical availability* of a machine being used if it is the case that the machine consists of many electrical or electronic subsystems?

The further considerations appear correct. But the problem is subtler. If a machine standing in a queue is switched off this time it can be counted as out of work and obviously out of repair. But if the machine engine is still on some subsystems of this machine are still in work state.

Generally: easy to say, harder to count.

Some further citations can be given here but they can only serve to create more disorder.

The situation needs to be clarified.

Availability, similarly to, e.g. reliability, maintainability, durability, **is an object ability** (property, feature) to fulfil (realize) its functions if it is needed. Availability belongs to the reliability scope of considerations, although this property is a point of consideration also in some other fields of science such as cybernetics, the theory of system efficiency (Figurski 1987, Sienkiewicz 1987) or telecommunications for instance.

The PN-ISO/IEC 2382-14: 2001 standard gives the following definition of availability:

'the ability of an object to be in a state to perform a required function under given conditions at a given instant time or over a given interval, assuming that the required external resources are provided.'

Additionally, the following comment is given:

'The availability defined here is an intrinsic availability where external resources other than maintenance resources do not affect the availability of the functional unit. Operational availability, on the other hand, requires that the external recourses be provided.'

A similar definition can be found in BS3811. However, Federal Standard 1037C contains the information that 'the degree to which a system is operable and in a committable state at the start of a mission, when the mission is called for at an unknown, i.e. a random, time.'

Availability is the property of objects that can be repaired, i.e. the exploitation process from a reliability point of view is a two-state one: work—repair (upstate—downstate). In the absolute majority of cases, this process is the alternative one, although there are some special cases where this process is not alternative. Czaplicki (1985) considered such a case.

Similar to the case of reliability, several different measures for availability are defined.

Generally, from the mathematical point of view, **all measures of availability are probabilities**.

The first papers on availability appeared in the 1960s together with reliability considerations (Hosford 1959/1960, Bielka 1960, Malikov *et al.* 1960, for example). Nagy (1963), Bailey and Mikhail (1963), Gnedenko *et al.* (1965), Thompson (1966), and Gray and Lewis (1967) have treated the problem of estimation of availability from statistical data. Brender's (1968) two papers gave a comprehensive lecture on the definition of availability: steady-state availability, point availability, steady-state mission availability, steady-state repair availability, transient point availability, steady-state availability of the second kind, transient mission availability and steady-state repeated demand availability. Problems connected with availability were viewed as so vital that in 1971 alone a number of publications appeared, such as: Martz, Kodama *et al.* Nakagawa and Goel, Das, McNichols *et al.* A dozen years later the term *availability theory* was formulated (see, for instance, Baxter 1985).

The most important measure of availability is the probability of an event that the object is in work state (upstate) i.e.:

$$A(t) = P\{\Psi(t) = 1\}. \tag{16.1}$$

This measure is called *pointwise availability* (see Kodama and Sawa 1986 or Malada 2006, for instance); this term has been used since the late 1950s (Hosford 1959/1960). According to some Authors, an alternative term for this probability is *instantaneous availability* (Elsayed 1996, www.weibull.com 2007, for instance).

Define the process $\Psi(t)$. Looking at Figure 5.1, the following can be written:

$$\Psi(t) = \begin{cases} 1 & Z''_n < t \le Z'_{n+1} \\ 0 & Z''_{n+1} < t \le Z''_{n+1} \quad n = 0, 1, \dots . \end{cases} \tag{16.2}$$

Defining two additional functions connected with the process:

- The probability distribution of work times:

$$P(t_{pi} < t) = F(t) \quad i = 1, 2, \ldots \tag{16.3}$$

- The probability distribution of repair times:

$$P(t_{ni} < t) = G(t) \quad i = 1, 2, \ldots \tag{16.4}$$

It is now possible to find a formula describing the considered availability function for the process. It is given by the formula:

$$A(t) = 1 - F(t) + \int_0^t [1 - F(t - x)] dH_\Phi(x) \tag{16.5}$$

where:

$$H_\Phi(t) = \sum_{n=1}^{\infty} \Phi_n(t) = \sum_{n=1}^{\infty} P(Z_n'' < t) = \sum_{n=1}^{\infty} \int_0^t F_n(x - u) dG_n(u) \tag{16.6}$$

(see Kopociński 1973, pp. 284–285).

In mine practice, apart from a few special cases concerning the operation of emergency units such as rescue systems (readiness systems), the availability of an object at a particular moment in time is not important.

The very wide application in engineering practice and mining practice has also found the limited value of the function $A(t)$ called **steady-state availability** or **long-run availability** or **limiting availability** (Gnedenko *et al.* 1969, Ryabinin 1976, Kiliński 1976, Adachi *et al.* 1979, Beichelt and Fischer 1979, wikipedia 2007, for instance). This parameter is defined by the well-known formula:

$$A = \lim_{t \to \infty} A(t) = \frac{E(t_p)}{E(T_p) + E(T_n)} \tag{16.7}$$

In many engineering publications and elaborations this measure used to be known in short as availability (or sometimes, unfortunately, *mechanical availability*).

Notice that this parameter is still the probability. To put it precisely, formula (16.7) determines **the probability of an event that the object is in work state at any moment of time**. This measure is not connected with a precise moment t at all.

This is one of the most important reliability/availability parameters of repairable technical objects. Its estimate is usually given by the producer on the list of the main parameters of its product. This information is very useful for estimating object production cycles, and, later, estimating object effective output and other measures of object efficiency of exploitation/operation.

The exploitation process of both types of machines modelled in this paper are now considered. First, the shovel is considered.

Generally, each exploitation process of a technical object has rich contents. The problem of how it should be determined depends on the answer to the question: why is this description being made? Or in other words, what is the purpose of its construction? For the considerations contained in this paper, it is enough for the exploitation process $\Omega(t)$ of the shovel to be identified as a three-state one. Such a process is usually presented as the function as shown on Figure 16.1 or in chart form in Figure 16.2.

The presentation of the process in graph form (*vide*: graph theory) gives more information. Alongside the possible states inventory there is information on possible transitions between states, and the intensities of these passages are also included. This is interesting from the exploitation point of view similarly to basic exploitation parameters such as probability that a machine is in a given state, intensity of transition between states, mean time of each state, etc. From a reliability point of view, the two states $\mathfrak{K}_{nd}$ and $\mathfrak{K}_{zd}$ can be lumped to give one state—the work state of the shovel:

$$\mathfrak{K}_p = \mathfrak{K}_{nd} \cup \mathfrak{K}_{zd}.$$

The process, $\Omega(t)$, is then converted into a two-state one, $\Psi(t)$. The availability of the shovel can easily be determined.

If the exploitation process of the truck is now considered, then one more state has to be incorporated—reserve (Figure 16.3).

Observe that the exploitation process shown in the graph's form carries information not only on transitions between states characteristic of trucks but also because of the existence of the whole

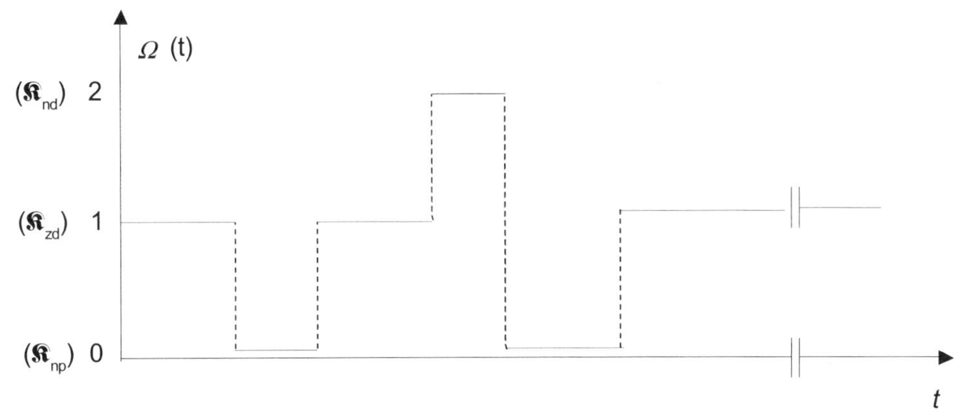

Figure 16.1. Process $\Omega(t)$ of changes of states: accessibility for loading—repair—inaccessibility for loading for shovel in the classical mathematical form, Ω(t) shown as the function key words: $\mathfrak{K}_{nd}$—state of inaccessibility for loading, $\mathfrak{K}_{zd}$—state of ability (and also accessibility) for loading, $\mathfrak{K}_{np}$—repair state.

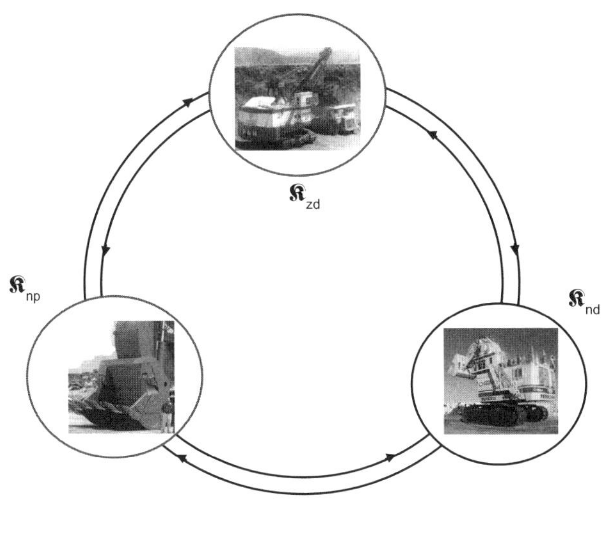

Figure 16.2. Process $\Omega(t)$ of changes of states in the graphs form: accessibility for loading—repair—inaccessibility for loading for shovel.

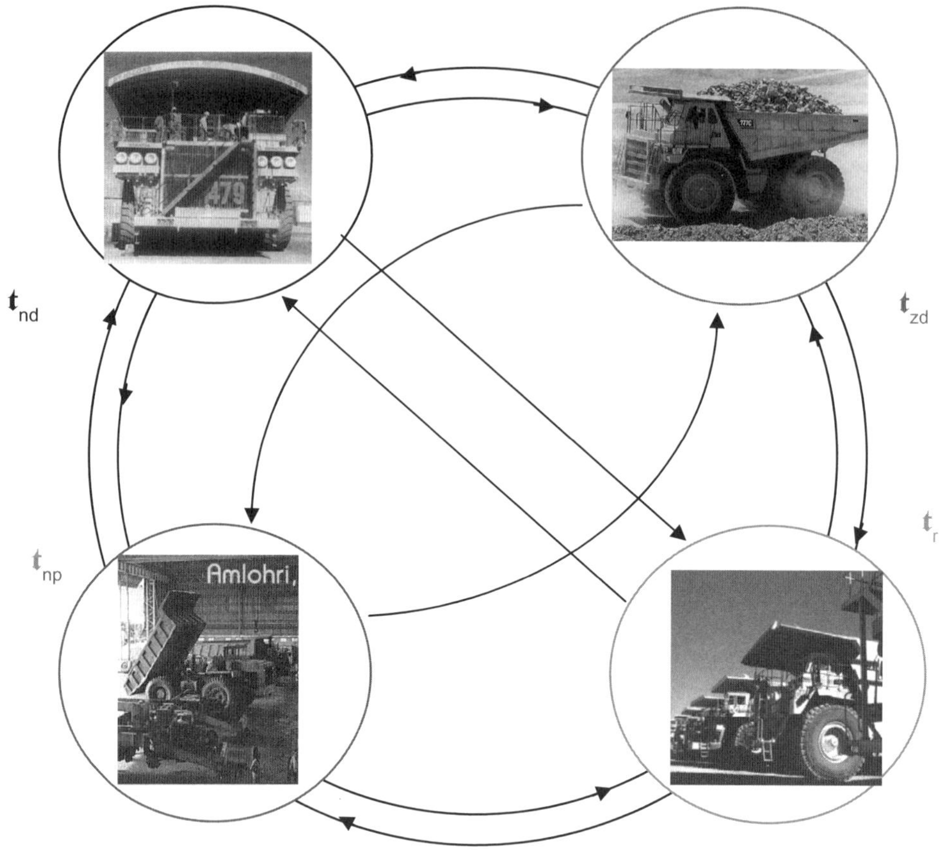

Figure 16.3. Process $\Lambda(t)$ of changes of states in the graphs form for trucks: accessibility for transporting—repair—inaccessibility for transporting—reserve.

system: the shovel-truck one. The transition from work to reserve appears when a certain shovel is down and the truck dispatcher decides to withdraw some trucks to reserve.

In order to obtain the possibility of accessing the truck availability, states t_{nd} and t_{zd} need to be lumped and state t_r needs to be extracted, that is the process needs to be transformed

$$\Lambda(t): < \mathfrak{S}; \lambda_{ij}; \mathbf{t} > , \mathfrak{S} : < \mathbf{t}_{zd}, \mathbf{t}_{nd}, \mathbf{t}_{np}, \mathbf{t}_{r} >,$$

where λ_{ij} are transitions between states

to

$$< \mathfrak{S} - \mathbf{t}_r, \mathbf{t}_{zd} \cup \mathbf{t}_{nd}; \lambda_{ij}; \mathbf{t} >$$

obtaining the process $\Psi(t)$.

The situation changes when the object considered instead of an element becomes a system. Instead of a 0, 1 situation—upstate, downstate—there is a multiple-valued one—partly upstate, partly downstate. Therefore, gradation of availability is introduced—a system is available to a certain extent. Availability measures are functions of the structure of system as well as the actual state of system. For continuous systems, availability can be calculated from the matrix of transitions between states (e.g. Czaplicki and Lutyński 1987 pp. 103–112). This matrix is the basis for the calculation of the steady-state probabilities for the exploitation process considered. These probabilities can be used to calculate basic system parameters, among other things the system output. For queuing models, the availability can be calculated based on the limited probability distribution states of the system (e.g. Czaplicki 2004 p. 36 Maryanovitch model).

This cyclic system is in a more convenient situation because trucks work in parallel from the reliability point of view, thus the mean value can be used. For this reason the truck steady-state availability in the system A_{ts} can be evaluated as

$$A_{ts} = 1 - \frac{E(X)}{m + r} \tag{16.8}$$

where *E*(*X*) is obviously the expected number of failed trucks in the system. This mean can be calculated using formula (5.1) because the probability density function is given in the continuous form.

Thus, the formula for this expected number can be presented. Looking at expression (7.13), the following can be written:

$$E(X) = \int_0^r xK_4 g_4(x)dx + \int_r^k xK_5 g_5(x)dx + \int_k^{m+r} xK_6 g_6(x)dx \tag{16.9}$$

CHAPTER 17

Final remarks

This book aimed to conduct a comprehensive analysis of the exploitation process of shovel-truck systems by modelling the required issues in the appropriate sequence. In this way, a specific procedure was created. The following recalls these problems one by one.

1. Analyze the reliability and accessibility of shovels
2. Discuss the functioning of a truck-repair shop system.
   The following items were considered here: the reliability of trucks, existence of haulers reserve, repair shop service capability, possibility of occurrence of a queue of failed trucks at the shop;
   Extensive analysis was also made investigating how the basic system parameters change when input data varies
3. Verify the *goodness* of the selection of structural truck system parameters: proper selection of the number of trucks directed to work, number of trucks in reserve and number of repair stands
4. Consider the reliability of repair stands
5. Construct the probability distribution of a number of trucks in work state
6. Discuss the functioning of a shovel-truck system
7. Construct a set of conditional efficiency measures for the system
8. Analyze truck-dispatching decisions and their repercussions
9. Indicate what kind of probability measures determine the quality of decisions made by a dispatcher
10. Construct a set of unconditional efficiency measures for the system
11. Outline system productivity measures
12. Look at the influence of spare loaders for system efficiency.

These 12 points create the whole procedure.

Moreover, two additional topics were discussed that are vital for mine practice. These are the:

a. Influence of implementation of priority in truck dispatching; priority concerning type of rock extracted removal
b. Influence of increasing the time travelled by a truck on the system parameters.

The book also aimed to provide a penetrative analysis of 6 stochastic mechanisms, namely the:

1. Influence of reliability and accessibility of power shovels on system efficiency and decisions made by a truck dispatcher
2. Influence of reliability of hauling machines on the number of repair stands needed and number of trucks needed
3. Influence of truck accessibility on the number of haulers and repair stands needed
4. Influence of reliability of repair stands on the number of failed trucks in a queue waiting for repair, on the number of trucks in work state and the number of repair stands needed
5. Influence of all these properties on the number of trucks at power shovels, system efficiency and productivity
6. Influence of implementation of type of priority in truck dispatching on system performance parameters.

The main characteristic feature of the considerations presented was the continual interlacing of theory with practice. Difficult, sometimes sophisticated mathematical methods were saturated by practical details.

It is possible to make use of some parts of the considerations presented in this monograph during the design process of shovel-truck systems. The main difference between the investigations made here and the material needed for design hinges on the necessity of replacement of the parameters given during discussion here by predicted random variables of—sometimes unknown—probability distributions.

To conclude these considerations it is worth noting that this procedure has three important properties allowing for the proper results of considerations to be achieved.

First, several probability distributions are frequently calculated. To check whether calculations are correct, the results can be verified by computing the sum of probabilities that should be close to unity.

The second property is the construction of mathematical expressions employed by diffusion approximation. Power exponents in formulas (2.4) and (2.5) and in their further modifications are very often high and sensitive to any changes. Even a small error in a value gives a bad result. This indicates that something is wrong.

The last property is that final outcomes are easy to understand and verify with practice, e.g. measures of the productivity of the system. If something is wrong in the procedure, the results will be nonsensical.

The author is well aware that much can still be done to improve this procedure, and will be very grateful for any comments concerning the methods used.

References

Adachi, K., Kodama, M. and Ohashi M., 1979. K-out-of-n: G system with simultaneous failure and three repair policies. *Microelectronics and Reliability*. Vol. 19, 4: 351–361.

Adamkiewicz, W., 1982. *Introduction to rational utilization of technical objects*. WkiŁ, Warsaw. (In Polish).

ARINC 1964. *Research Corporation: Reliability Engineering*. Englewood Cliffs. N.J., Prentice-Hall.

Arndt, U. and Franken, P., 1979. Непрерывность обобщенных регенерующих процессоb. Изb. AH CCCP, Сер. Техн. Кибернетика 3: 94–97.

Ataee-pour, M., Damiri, A. and Sadri, A., 2005. Optimization of the haulage system in Choghart iron ore mine using ARENA. *20th World Mining Congress*, Teheran, Iran, 7–11 November: 225–230.

Atis Telecom Glossary, 2000.

Aurignac, R., Engel, F. and Ferry, D., 1968. Etude du roulage general d'une exploitation miniere par simulation sur ordinateur. *Revue de l'Industrie Minerale*, Octobre.

Awuah-Offei, Kwame, Temeng, V.A. and Al-Hassan, S., 2003. Predicting equipment requirements using SIMAN simulation—a case study. *Mining Technology. Trans. Inst. Min. Metall. A*. Vol. 112, No 3: 180–184.

Bailey, J.H. and Mikhail, W.F., 1963. Sequential testing of electronic systems. *Proc. Ann. Symposium on Reliability*: 391–400.

Barbaro, R.W. and Rosenshine, M., 1986. *Evaluating the productivity of shovel-truck materials haulage system using a cyclic queuing model*. SME Fall Meeting, St. Louis, Missouri, Sept. 7–10.

Barnes, R.J., King, M.S. and Johnson, T.B., 1979. Probability techniques for analysing open pit production systems. *Proceedings of the 16th APCOM*, O'Neil T., (ed.) University of Arizona, Tucson: 462–476.

Bascetin, A., 2004. An application of the analytic hierarchy process in equipment selection at Orhaneli open pit coal mine. *Mining Technology. Trans. Inst. Min. Metall. A*. Vol. 113, No 3: 192–199.

Bauer, A. and Calder, P., 1973. *Planning Open Pit Mining Operations Using Simulation*. APCOM, Johannesburg, South African Institute of Mining and Metallurgy: 273–298.

Baxter, L.A., 1985. Some notes on availability theory. *Microelectronics & Reliability*. R-25, 5: 921–926.

Beichelt, F. and Fischer, K., 1979. On a basic equation of reliability theory. *Microelectronics and Reliability*. Vol. 19, No 4: 367–370.

Belak, S., 2005. Terotechnology. *Oceans '04. MTS. IEEE Techno-Ocean '04*. Vol. 1: 201–204.

Berman, M., 1978. Regenerative multivariate point processes. *Adv. Appl. Prob.* 10, 2: 411–430.

Będkowski, L. and Dąbrowski, T., 2006. *Exploitation foundation. Part II. Exploitation reliability foundations*. WAT, Warsaw. (In Polish).

Bhaudury, B. and Basu, S.K., 2003. *Terotechnology: Reliability engineering and maintenance management*. Asian Books.

Bishele, I.W., Gojzman, E.I. and Saratowski E.G., 1964. Имитрущая матиматическая модель работы рудничного транспорта. ИГД Скочинского, Москва.

Bielka, R. P., 1960. Availability—a system function. *IRE Trans. Reliability and Quality Control*, Vol. RQC-9, September: 38–42.

Birolini, A., 1971. Some applications of regenerative stochastic processes to reliability theory—part one: Tutorial introduction. *IEEE Trans. on Reliability*: 186–194.

Bonates, E., 1992. *The development of assignment procedures for semi-automated truck/shovel systems*. Ph. D. Thesis. Dept. of Mining & Metallurgical Eng., McGill University, Montreal, Canada, March.

Bonates, E., 1993. The development of assignment procedures for truck dispatching. *Proceedings of the 2nd International Symposium on Mine Mechanization and Automation*. Luleå, Sweden, 7–10 June: 543–553.

Borozdin, O.P. and Jeżow I. I., 1976. Сильно регенерующие случайные процессы. Докл. AH УССР, сер. A, 9: 771–773.

Bozorgebrahimi, A., Hall, R.A. and Morin, M.A., 2005. Equipment size effects on open pit mining performance. *International Journal of Surface Mining, Reclamation and Environment*. Vol. 19, No 1: 41–56.

Bozorgebrahimi, E., Hall, R.A. and Blackwell, G.H., 2003. Sizing equipment for open pit mining—a review of critical parameters. *Mining Technology*. Trans. Inst. Min. Metall. A. Vol. 112: A171–A179.

Brender, D.M., 1968a. The prediction and measurement of system availability: a Bayesian treatment. *IEEE Trans. on Reliability*, R-17, 3: 127–137.

Brender, D.M., 1968b. The Bayesian assessment of system availability: advanced application and techniques. *IEEE Trans. on Reliability*, R-17, 3: 137–146.

BS3811: 1993. *Glossary of terms used in Terotechnology*.

Bucklen, E.P., Suboleski, S.C., Preklaz, L.J. and Lucas J.R., 1968. Computer applications in underground mining systems. Vol. 4, BELTSIM *Program, Research Development Report No 37*, Virginia Polytechnic Institute.

Cabrera, J.G. and Maher, M.J., 1973. Optimizing earthmoving plant: solution for the excavator-trucks system. *Highway Research Record*. No. 454: 7–15.

Callut, J. and Dupont, P., 2004. A Markovian approach to the induction of regular string distributions. *7th Intern. Colloquium on Grammatical Inference*. Athens. Greece.

Carmichael, D.G., 1987. *Engineering queues in construction and mining*. John Wiley & Sons.

Caterpillar Tractor Co. *Handbook of earthmoving*. 1981.

Caterpillar performance handbook. Peoria, Illinois, 1996.

Chan, A.W., 1982a. An interactive simulation model of conveyor systems in oil sands mining. *NRC Report LTR-CS-227*, February.

Chan, A.W., 1982b. An interactive simulation model of bin/hopper systems in oil sands mining. *NRC Report LTR-CS-231*, August.

Chan, A.W., 1982c. An interactive simulation model of conveyor systems. *CIM Bulletin*. Vol. 847. November: 81–83.

Church, H.K., 1981. *Excavation handbook*. McGraw-Hill Book Co.

Connell, J.P., 1969. *Determination of the fleet availability using probability calculations*. Preprint 69–AR-46, SME/AIME, New York, February.

Connell, J.P., 1973. In *SME Mining engineering handbook*. Vol. 2: 18–16: 18–33.

Cross, B. and Williamson, G., 1969. *Digital simulation of an open-pit truck haulage system*. APCOM.

Czaplicki, J.M., 1985. A certain model of reliability diagnostic of renewal objects. *Microelectronics & Reliability*, 4: 695–698.

Czaplicki, J., and Lutyński, A., 1987. Horizontal transport. *Reliability problems*. Silesian University of Technology. Textbook No 1330, Gliwice.

Czaplicki, J.M., 1988. On availability for equipment selection. *Proceedings of the International Symposium on Mine Planning and Equipment Selection*, Calgary, November 3–4: 261–265.

Czaplicki, J.M., 1989a. *Notes from reliability investigation of the Nchanga Open Pit 1986–1988*, Copper belt. Mining Engineering Dept. of University of Zambia, Lusaka (unpublished).

Czaplicki, J.M., 1989b. Towards a better description of shovel-truck systems. *International Symposium on Off-Highway Haulage in Surface Mines*, Edmonton 15-17 May: 69-73.

Czaplicki, J.M., and Temeng, V. A., 1989. Reliability investigation of a shovel-truck system at the Nchanga open-pit. Proceedings of the International Symposium on Off-Highway Haulage in Surface Mines. Edmonton, 15–17 May: 105–110.

Czaplicki, J.M., 1990a. On availability of a piece of equipment. *Intern. Symposium on Mine Planning and Equipment Selection*, Calgary, 7–9 November: 209–216.

Czaplicki, J.M., 1990b. A simple method for the estimation of shovel-truck system efficiency. *Mineral Resources Engineering*, vol.3, 1: 21–29.

Czaplicki, J.M., 1990c. The use of the *G/G/1* queuing model for simulation of heavy traffic flows for truck-shovel systems in open pit mines. *Mineral Resources Engineering*, vol. 3, 3: 211–219.

Czaplicki, J.M., 1992a. Probability distribution of a working truck number per one working shovel in open pit mining. *Journal of Mining Research*, vol.1, No 3: 31–38.

Czaplicki, J.M., 1992b. On a number of spare haulage units for the shovel-truck system in open pit mines. *International Journal of Surface Mining and Reclamation*, 6: 57–59.

Czaplicki, J.M., 1993a. Some reliability problems in the design of systems shovel—truck of surface mining. *XXI Reliability Winter School*. Szczyrk, 5–8 January: 177–184. (in Polish).

Czaplicki, J.M., 1993b. On selection of number of trucks for the shovel—truck system of surface mining. *VII Domestic Symposium on Terotechnology*. Radom-Kozubnik 1993: 93–101. (in Polish)

Czaplicki, J.M., 1993c. Equipment selection for aggregate stone loading—a case of study. *Proceedings of the International Conference on Designing, Production and Exploitation of Mining Equipment and Automation Devices*, Gliwice 16 September: 343–349.

Czaplicki, J.M., 1994. Some aspects of truck reserve in the system of open pit mining. *XXII Reliability Winter School*, SPE KBM PAN, Szczyrk: 23-33. (in Polish).

Czaplicki, J.M., 1997. Productivity of the shovel-truck system based on an analysis of different models. *Mineral Resources Engineering*. Vol. 6, No. 3: 117–128.

Czaplicki, J.M., 1999. A new method of truck number calculation for shovel-truck system. *Mineral Resources Engineering*. Vol. 8, No.4: 391–404.

Czaplicki, J.M., 2000. Selection of the predicted FEL system to achieve the loading task. *International Scientific Conference*, Ostrava, 5–7 September. (CD)

Czaplicki, J.M., 2002. Simple method of large shovel-truck system calculation. *Górnictwo Odkrywkowe* (*Surface Mining*) 4: 24–29.

Czaplicki, J.M., 2004a. Method of shovel-truck-crusher-conveyor system calculation. *5th International Symposium on Mining Science and Technology*, Xuzhou, Jiangsu, P.R., China, 20–22 October: 787–792.

Czaplicki, J.M., 2004/2005. Method of "shovel-truck-inclined hoist" system calculation. *Górnictwo Odkrywkowe* (*Surface Mining*), No. 7–8: 71–75.

Czaplicki, J.M., 2004b. *Elements of theory and practice of cyclic systems in mining and earthmoving*. Silesian University of Technology. Textbook. Gliwice. (In Polish)

Czaplicki, J.M., 2005. On the system: "stream of mineral—shaft bin—hoist" in relation to the classical design of hoist installation. *Proceedings of the International Conference "Shaft Transport 2005"*. Zakopane 21–23 September: 17–31. (In Polish).

Czaplicki, J.M., 2006. *Modeling of the exploitation process of shovel-truck system*. Papers of Silesian University of Technology. 1740. Gliwice. (In Polish).

Darnell, H., 1979. Guest Editorial. *Terotechnica*. Vol. 1, No. 1.

Das, P., 1971. Point-wise availability of an electronic system with reduced efficiency class of components. *Microelectronics and Reliability*. Vol. 10, 2: 61–66.

DIN 40041: *Bergriffsbestimmungen zur Zuverlässigkeit elektrischer Bauelemente*.

Dohm, G.C., Jr., 1992. Open pit feasibility studies. In *SME MEH*.

Douglas, J., 1964. *Prediction of shovel-truck production: A reconciliation of computer and conventional estimates*. Technical Report No. 37. Department of Civil Engineering. Stanford University.

Downarowicz, O., 1997. *Exploitation system. Management of technical resources*. Gdańsk-Radom. (In Polish).

Dudczak, A., 2000. *Shovels. Theory and design*. PWN, Warsaw. (In Polish).

Eichler, P., 1968. Die Kostenberechnung der gleisgebundenen Hauptstrecken—forderung auf digitalen Rechenautomaten als Hilfmittel der Planung und Uberwachung. *Schlagel und Eisen*, 6.

Elbrond, J., 1977. *Calculation of an open pit operation capacity*. SME Fall Meeting and Exhibit, St. Louis, Missouri, 19–21 October.

Elbrond, J., 1990. Haulage systems analysis: Queuing theory. In *Surface Mining*. SME: 741–748.

Elsayed, E.A., 1996, *Reliability engineering*. Addison-Wesley.

Erarslan, K., 2005. Modeling performance and retarded chart of off-highway trucks by cubic spines for cycle time estimation. *Mining Technology*. Trans. Inst. Min. Metall. A. Vol. 114, No 3: 161–166.

Fabian, J., 1989. Cyclic mining systems in Czechoslovakian surface mines. *Proceedings of the International Symposium on Off-Highway Haulage in Surface Mines*, Edmonton 15–17 May: 205–209.

Federal Standard 1037 C.

Feller, W., 1949. Fluctuation theory of recurrent events. *Trans. Amer. Math. Soc.* 67, 1: 98–119.

Figurski, J., 1987. Machinery stock cybernetics. *Ossolineum*. (In Polish).

Firganek, B. and Wianecki, A., 1970. Examples of application of the simulation to analyze mine transport systems. *Design Problems*, 9. (In Polish).

Frimpong, S., 1986. *Productivity and economic analyses of the Apinto mechanization project*. Postgraduate Diploma Thesis. UST School of Mines, Ghana.

Frimpong, S., Whiting, J.M. and Szymański, J., 1997. Rigorous functional model for mine production equipment maintenance. *Proceedings of the International Symposium on Mine Planning and Equipment Selection*, Ostrava, Czech Republic, 3–6 September: 39–46.

Gaarslev, A., 1969. Stochastic models to estimate the production of material handling systems in the construction industry. *Technical Report No. 111*. The Construction Institute. Stanford University. August.

Gertzbah, I. B. and Kordonsky H. B. (Герцбах, И. Б., Кордонский, Х. Б.), 1966, Модели отказов. Изд. Советское Радио, Москва.

Gnedenko, B.W., Bielajev, J.K. and Soloviev, A.D., 1965. Математические методы в теории надёжности. Изд. Наука, Moscov.

English translation: Gnedenko, B.V., Belayev, Yu.K., and Solovyev, A.D., 1969. *Mathematical methods of Reliability theory*. Academic Press. New York.

Gilewicz, P., and Woof, M., 2005. Pushing tin—observations on the 2004 mining truck market. *World Mining Equipment*, March, vol. 29, No 2: 12–14.
Glinski, G.S., de Mercado, J.B. and Thompson, P.M., 1969. A diffusion method for reliability prediction. *IEEE Trans. of Reliability*. Vol. R-18, 4: 149–156.
Golosiński, T.S., 1991. Practical aspects of availability and reliability management in a large open pit mine. *Reliability Production and Control in Coal Mines*, Wollongong, 2–6 September: 127–130.
Gould, R.J., 1988. *Graph theory*. Benjamin—Cummings Pub. Co.
Graff, G., 1971. Simple queuing theory saves unnecessary equipment. *Industrial Engineering*, February: 15–18.
Gray, H.L. and Lewis, T.O., 1967. A confidence interval for the availability ratio. *Technometrics*. Vol. 9, August: 465–471.
Griffin, W., 1989. Perturbation theory applied to truck-shovel system simulation. *Proceedings of the International Symposium on Off-Highway Haulage in Surface Mines*, Edmonton 15–17 May: 83–87.
Griffin, W.H. and Szymanski, J.K., 1988. A simulation analysis model for continuous surface mine and trucks operation using SLAM II. *2nd Continuous Surface Mining Symposium*, Austin, Texas.
Gross, D. and Harris, C.M., 1974. Fundamentals of Queuing Theory. John Wiley: 214–217, 306–308.
Hagenbuch, L.G., 1987. Mining truck dispatching: The basic factors involved and putting these factors into mathematical formulas (algorithms). *International Journal of Surface Mining*. 1: 105–129.
Halachmi, B. and Franta, W.R., 1978. A diffusion approximation to the multi-server queue. *Management Science*. Vol. 24, No. 5: 552–559.
Harvey, P.R., 1964. Analysis of production capabilities. APCOM, *Colorado School of Mines Quarterly*.
Hartman, H.L., 1987. *Introductory mining engineering*. J. Wiley & Sons, Inc.
Haryono and Sivazlian, B.D., 1985. Analysis of the machine repair problem: a diffusion process approach. *Mathematics and Computer in Simulation*. Vol. 27: 339–364.
Hays, R.M., 1990. Trucks. In *Surface Mining. SME*: 672–690.
Heyman, D.P., 1975. A diffusion model application for the *GI/G/1* queue in heavy traffic. *The Bell System Technical Journal*. Vol. 54: 1637–1646.
Hewgill, J.C. and Parkes, D., 1979. Terotechnology—philosophy and concept. *Terotechnica*. Vol. 1, No. 1: 12–19.
Hosford, J.E., 1960. Measures of dependability. *Operation Research*. Vol. 8, No 1: 204–206.
Hufford, G., Griffin, W.H. and Sturgul, J.R., 1981. *An application of queuing theory to equipment selection for a large open pit copper mine*. SME-AIME Meeting, Denver, Colorado, 18–20 November.
Huk, J. and Łukaszewicz, J., 1973. Cyclic systems of mass servicing. *Applied Mathematics*, III, I: 85–104. (In Polish).
Humphrey, J.D., 1990. Walking draglines. In *Surface Mining SME*: 638–655.
Iglehart, D.L., 1965. Limiting diffusion approximation for the many server queue and the repair problem. *Journal of Applied Probability*. Vol. 2: 429–441.
Investopedia.
Jastriebienieckij, M.A., 1969. Об одном классе регенерующих случайных процессоb. Изb. АН СССР, Сер. Техн. Кибернетика 5: 50–60.
Juckett, J.E., 1969. A coal mine belt system design simulation. *3rd Conference on the Application of Simulation*. Los Angeles.
Kabak, I. W., 1969. System availability and some design applications. *Journal Oper. Research Soc. Amer*. Vol. 17, 5: 827–837.
Kammerer, B.A., 1988. In-pit crushing and conveying system at Bingham Canyon Mine. *International Journal of Surface Mining*. Vol. 2, No. 3: 143–147.
Kapliński, O., 1974. An application of the queue theory to optimize technological system. *Statistical Survey*, R. XXI. z. 4: 631–646. (In Polish).
Kaźmierczak, J., 2000. *Exploitation of technical systems*. Silesian University of Technology, Gliwice. (In Polish).
Kendall, D.G., 1953. Stochastic processes occurring in the theory of queues and their analysis by the method of the imbedded Markov chain. *Ann. Math. Statist*. 24: 338–354.
Kennedy, B.A., (ed.) 1990. *Surface Mining*. SME.
Kiliński, A., 1976. Industrial accomplishment processes. *Principles of theory. WNT*, Warsaw. (In Polish).
Kingman, J.F.C., 1965. The heavy traffic approximation in the theory of queues. *Proceedings of Symposium on Congestion Theory*. University of North Carolina Press, Chapel Hill, NC: 137–159.
Kobayashi, H., 1974. Application of the diffusion approximation to queuing network I: equilibrium queue distributions. *Journal of the Association for Computing Machinery*. Vol. 21, 2: 316–328.

Kodama, M., Fukuta, J. and Takamatsu, S., 1971. Mission availability a 1-unit system with allowed downtime. *IEEE Transactions on Reliability*: 268–270.

Kodama, M. and Sawa, I., 1986. Reliability and maintainability of a multicomponent series-parallel system with simultaneous failure under preemptive repeat repair disciplines. *Microelectronics and Reliability*, 1: 163–181.

Koenigsberg, E., 1958. Cyclic queues. *Operational Research Quarterly*, 9, 1: 22–35.

Kolonya, B., Stanic, R. and Hamovic, J., 2003. Simulation of in-pit crushing systems using AutoMod. *19th Mining Congress*, 1–5 November, New Delhi: 517–531.

Kopocińska, I., 1968. Mathematical models of cyclic queues. *Inwestycje i Budownictwo*. 18, Nr 12: 22–25. (In Polish).

Kopociński, B., 1973. *An outline of renewal and reliability theory*. PWN, Warsaw. (In Polish).

Korak, J. and Müller, G., 1987. Semi-mobile crushers for reducing hard overburden to beltable size at the Mae Moh Lignite Mine—A state-of-the-art report. *Proceedings of an International Symposium on Continuous Surface Mining*. Edmonton September 29—October 1: 595–603.

Kozioł, W. and Uberman, R., 1994. *Technology and structure of surface mine transport*. Wyd. AGH, Cracow. (In Polish).

Kozlov, B. and Ushakov, I. (Козлов, Б., and Ушаков, И.), 1966. Справочник по расчету надёЖности. Москва, Изд. Советское Радио

Kumar, K. and Sinha, D.K., 1989. Equipment operational reliability and availability in mineral haulage: A case of study. *Proceedings of the International Symposium on Off-Highway Haulage in Surface Mines*, Edmonton 15–17 May: 197–203.

Kumar, U., 1996. Maintenance strategy for mechanized and automated mining systems. *Proceedings of the International Symposium on Mine Planning and Equipment Selection*, Sao Paulo, Brasil: 553–558.

Kuruppu, M.D., 2004. New technologies available to maximizing equipment reliability. *Mine Planning and Equipment Selection*, Wrocław, 1–3 September: 455–459.

Lill, J.W., 1986. *Operating features and control aspects of the Palabora copper open pit*. The Planning and Operation of Open-Pit and Strip Mines. Deetlefs. J.P. Johannesburg, SAIMM: 1616–174.

Lizotte, Y. and Bonates, E., 1986. *Truck/shovel dispatching rules assessment using simulation*. SME Fall Meeting, St. Louis, Missouri, 7–10 September.

Lizotte, Y., Bonates, E. and Leclerc, A., 1989. Analysis of truck dispatching with dynamic heuristic procedures. *International Symposium on Off-Highway Haulage in Surface Mines*, Edmonton 15–17 May: 47–55.

Madge, D.N., 1964. Simulation of truck movement in an open pit mining operation. *Canadian Operational Research Society*: 32–41.

Maher, M.J. and Cabrera, J.G., 1973. The transient behavior of a cyclic queue. *Operational Research Quarterly*, 24. 4: 603–613.

Maher, M.J. and Cabrera, J.G., 1975. A multi-stage cyclic queuing model. *International Journal of Production Research*. 13, 3: 255–264.

Malada, A., 2006. *Stochastic reliability modeling for complex systems*. Ph. D. dissertation, Faculty of Engineering University of Pretoria.

Malikov, I.M., Polovko, A.M., Romanov, N.A. and Tchukreev, P.A., 1960. Основы теории и расчета надёжности. Судпромгиз.

Manula, C. and Rivell, R., 1974. A master design simulator. APCOM 12, *Colorado School of Mines Quarterly*.

Markovsky, J and Trcka, N., 2006. Lumping Markov chains with silent steps. *Proc. of 3rd Intern. Conference on the Quantitative Evaluation of Systems*: 221–232.

Martz, H.F., 1971. On single cycle availability. *IEEE Transactions on Reliability*. R-20, May: 21–23.

Maryanovitch T. P. (Марьянович, Т. П.), 1961, Надійнiсть системы зі змішаним резервом. Дон. АН УРСР, 8.

McDonald, D., 1978. On semi-Markov and semi regenerative processes II. *Ann. Prob.* 6, 6: 995–1014.

McNichols, R. J. and Messer, G. H., 1971. A cost-based availability allocation algorithm. R-20/3, 178–182.

Merris, R., 2000. *Graph theory*. Wiley Interscience.

Migdalski, J. (ed.), 1982. *Reliability Handbook. Mathematical Principles*. WEMA. Warsaw. (In Polish).

Mining Engineering Handbook. Vol. 4, Śląsk, Katowice 1982. (In Polish).

Morgan, W.C. and Peterson L.L., 1968. Determining shovel-truck productivity. *Mining Engineering*. Vol. 20, No.12: 76–78.

Mrig, G.C., 1991. *Reliability of machines and systems used in production process in mines—A case study of South Eastern Coalfields Limited, India*. Reliability Production and Control in Coal Mines, Wollongong 2–6 September: 104–110.

MSN dictionary.
Nagy, G., 1963. The reliability of repairable systems. *Proc. Ann. Symposium on Reliability*: 93–108.
Nakagawa, T. and Goel, A.L., 1971. A note on availability for a finite interval. *IEEE Transactions on Reliability*: 271–272.
Nanda, N.K., 2003. *Optimization of mine production system through Operation research Techniques*. 19th Mining Congress, 1–5 November, New Delhi: 582–595.
Nenonen, L.K., 1982. Interactive computer modeling of truck haulage systems. *CIM Bulletin*. Vol. 75, No. 847, November: 84–90.
Nenonen, L.K., Graefe, P.W.U. and Chan, A.W., 1981. *Interactive computer modeling of truck/shovel operations in an open-pit mine*. Winter Simulation Conference: 133–139.
Newhart, D.D., 1977. *Bethbelt-1, A belt haulage simulator for coal mine planning*. Bethlehem, Pennsylvania. Research Report File 1720–2.
O'Shea, J.B., Slutkin, G.N. and Shaffer, L.R. 1964. An application of the theory of queues to the forecasting of shovel-truck fleet productions. *Civil Engineering Studies*. Construction Research Series No. 3. University of Illinois.
Oraee, K. and Asi, 2004. *Fuzzy model for truck allocation in surface mines*. Mine Planning and Equipment Selection, Wrocław 1–3 September: 585–591.
Palm, C., 1947. The distribution of repairmen in servicing automatic machines. *Industritidningen Norden*, 17: 75–80, 90–94, 119–123. (In Swedish).
Panagiotou, G.N., 1993. Optimizing the shovel-truck operation using simulation and queueing models. *Proceedings of the 2nd International Symposium on Mine Mechanization and Automation*. Luleå, Sweden, 7–10 June: 601–607.
Panagiotou, G.N. and Michalakopoulos, T.N., 1994. Analysis of shovel-truck operations using STRAPAC. *Proceedings of the International Symposium on Mine Planning and Equipment Selection*, Istanbul, Turkey, 18–20 October: 295–300.
Panagiotou, G.N., Michalakopoulos, T.N., and Kladis C. D., 1996. A Maintenance management system for mobile equipment. *Proceedings of the International Symposium on Mine Planning and Equipment Selection*, Sao Paulo, Brazil: 559–563.
PN-EN 13306: 2006. *Maintenance terminology*.
PN-ISO/IEC 2382–14: 2001. *Information technology—Vocabulary—Part 14—Reliability, maintainability and availability*.
Purohit, G.S. and Nanda, N.K., 1995. *Optimization of shovel dumper operation in an opencast mine by using queuing model—A case of study*. 5th National Seminar on Surface Mining, 4–5 December: 37–39.
Reliability in technique. Set of Polish standards. Alfa-Wero. 1997. (In Polish).
Rist, K., 1961. *The solution of a transportation problem by use of a Monte Carlo technique*. APCOM 1, Tucson, University of Arizona, March: D1–D18.
Russell, A.C., 2006. Hybrid system solves GPS problems at Morenci. *Engineering & Mining Journal*. May: pp. 46–52.
Ryabinin, I., 1976. Reliability of engineering systems. *Principles and Analysis*. Mir Publishers. Moscow.
Rykov, W. W. and Jastriebienieckiy, M.A., 1971. *О регенерующих процессах с несколькими типами точек регенерации. Кибернетика*, 3: 82–86.
Sajkiewicz, J., 1979. *Zarys teorii eksploatacji systemów maszynowych*. (An outline of theory of exploitation of machinery systems.) Wyd. Politechniki Wrocławskiej, Wrocław. (In Polish).
Sharma, D.K., 1989. Fleet imbalance analysis in large open pit coal mines in India. *Proceedings of the International Symposium on Off-Highway Haulage in Surface Mines*, Edmonton 15–17 May: 89–93.
Sharma, D.K. and Ekka, L., 1989. Shovel-truck productivity studies in open-pit copper mines. *Proceedings of the International Symposium on Off-Highway Haulage in Surface Mines*, Edmonton 15–17 May: 111–116.
Sienkiewicz, P., 1987. Theory of system effectiveness. *Ossolineum* 1987. (In Polish).
Sivazlian, B.D. and Wang, K. –H., 1988. Diffusion approximation to the G/G/R machine repair problem with warm standby spares. *Naval Research Logistics Quarterly*.
Sivazlian, B.D. and Wang, K.H., 1989. System characteristics and economic analysis of the G/G/R machine repair problem with warm standbys using diffusion approximation. *Microelectronics & Reliability*, 29, 5: 829–848.
SME. *Mining Engineering Handbook*. New York 1973, vol. 2.
SME. *Mining Engineering Handbook*. 1992, vol. 2.
Smith, W.L., 1955. Regenerative stochastic processes. *Proc. Roy. Soc. Ser. A*, 232: 6–31.
Smith, W. L., 1958. Renewal theory and its ramifications. *Journal Royal Stat. Soc.*, Ser. B. Vol. 20: 243–302.

Soumis, F., Ethier, J., McInnis, D. and Elbrond, J., 1986. *A new method for automatic truck dispatching in an open pit mine*. SME Fall Meeting, St. Louis, Missouri, Sept. 7–10.
Spaugh, J.M., 1962. The use of the theory of queues in optimal design of certain construction operations. *Technical Report of National Science Foundation* (NSF G 15933). Department of Civil Engineering. University of Illinois, Urbana. Illinois.
Stryszewski, M., 1981. Operation assessment and the output control of the system shovel—truck—crusher using the queue theory. *Papers of AGH, Mining*. Year 5, 2: 157–169. (In Polish).
Stuart, A. and Ord, J.K., 1998. Kendall's advanced theory of statistics. *Vol. 1: Distribution theory*.
Sturgul, J.R., 2000. *Mine design. Examples using simulation*. SME.
Szymański, J. K. and Srajer, V., 1989. *An integrated monitoring system and simulation analysis model for trucks operation using SLAM II*. International Symposium on Off-Highway Haulage in Surface Mines, Edmonton 15–17 May: 75–81.
Takács, L., 1962. *Introduction to the theory of queues*. New York, Oxford University Press.
Talbot, K., 1977. *Simulation of conveyor belt networks in coal mines*. APCOM 15, Brisbane, Australia.
Tattersall, A., 2004. Tire awareness program pays bug dividends at Coeur Rochester. *Engineering & Mining Journal*. February: 23–25.
Taylor, L.E., Benham, A.J., Brown, T.J., Lusty, P.A.J., Minchin, D.J., 2006. *World mineral production*. British Geological Survey, Natural Environment Research Council. Keyworth, Nottingham.
Teicholz, P., 1963. *A simulation approach to the selection of construction equipment*. Technical Report No. 26. Construction Institute. Stanford University. June.
Teicholz, P. and Douglas, J., 1963. *An analysis of two-link material handling systems with one carrier in one of the links*. Technical Report No. 29. Department of Civil Engineering. Stanford University.
Temeng, V. A., 1988. *Probabilistic analysis of reliability and effectiveness of elementary shovel-truck system at Nchanga Open-pit*. M. Sc. Dissertation, University of Zambia, Mining Engineering Dept.
Terotechnology Handbook. Her Majesty's Stationary Office. 1978.
Terex. 1981. *Production and cost estimating of material movement with earthmoving equipment*. Manual.
Thesaurus.com
Thomas, L.J., 1979. *An introduction to mining. Exploration, feasibility, extraction, rock mechanics*. Methuen of Australia.
Thompson, M., 1966. Lower confidence limits and test of hypotheses for system availability. *IEEE Trans. on Reliability*. Vol. R-15, 1: 32–36.
Tseng, Y.Y., 1973. *Queuing approach to the two-link materials handling system and its application in activities*. M. Eng. Thesis No. 578. Asian Institute of Technology. Bangkok.
Vergne, J., 2003. *Hard Rock Miner's Handbook*. McIntosh Engineering.
Walker, G., 1985. *Behold Behemoth: Toward the 1000 tonne truck*. Mining Equipment Selection Symposium. Calgary, University of Calgary/CANMET, November.
Wang, X. and Zhao, L., 1997. A study of haulage technology and equipment in large open pit mines. *Proceedings of the International Symposium on Mine Planning and Equipment Selection*, Ostrava, Czech Republic, 3–6 September: 511–514.
Weber, L., 2005. *Recent trends in the world's minerals supply*. 20th World Mining Congress, Teheran, Iran, 7–11 November: 473–480.
Weicheng, L.X.Z. and Youdi, Z., 2004. *Application of genetic algorithm to optimize the number and size of equipment of surface mine*. Mine Planning and Equipment Selection, Wrocław, 1–3 September: 637–639.
White, L., 2006. Precious Metals Continue to Boom. *Engineering & Mining Journal*. May: 42–44.
Wikipedia.
Woodrow, G.B., 1992. *Benchmarks of Performance for Truck and Loader Fleets*. 3rd Large Open Pit Mining Conference. Macay, 30.08–03.09: 119–125.
Wright, E.A., 1988. Truck dispatching on a personal computer. *Proceedings of the International Symposium on Mine Planning and Equipment Selection*, Calgary November 3–4: 429–433.
www.allmining.com/miningequipmentsales.: 2005 Surface Mining Equipment Sales Top US$ 3 billion.
www.buengobierno.com/admin/files/ElCerrejon.pdf
www.jsonline.com/story/index.aspx?id=382576.: Mining equipment makers hit pay dirt.
www.prweb.com/releases/2005/12/prweb324869.: 2006 Surface Truck Predictions, Musings & Questions.
www.weibull.com
www.wencommine.com/products_fleetmon
Yao, D. D., 1985. Refining the diffusion approximation for the *M/G/m* queue. *Operation Research*. Vol. 33, 6: 1266–1277.

Zakharin, A. M., 1972. Lumping the states of Markov and semi-Markov processes and the use of the lumping operation for the investigation of stochastic processes. *Journal of Cybernetics and System Analysis*. Vol. 8, 4. July: 589–598.

Zhongzhou, L. and Qining, L., 1988. Erlangian cyclic queuing model for shovel-truck haulage systems. *Proceedings of the International Symposium on Mine Planning and Equipment Selection*, Calgary November 3–4: 423–428.

Subject index